Hexavalent Chromium Removal Using Anion Exchange and Reduction With Coagulation and Filtration

Awwa Research Foundation

About the Awwa Research Foundation

The Awwa Research Foundation (AwwaRF) is a member-supported, international, nonprofit organization that sponsors research to enable water utilities, public health agencies, and other professionals to provide safe and affordable drinking water to consumers.

The Foundation's mission is to advance the science of water to improve the quality of life. To achieve this mission, the Foundation sponsors studies on all aspects of drinking water, including supply and resources, treatment, monitoring and analysis, distribution, management, and health effects. Funding for research is provided primarily by subscription payments from approximately 1,000 utilities, consulting firms, and manufacturers in North America and abroad. Additional funding comes from collaborative partnerships with other national and international organizations, allowing for resources to be leveraged, expertise to be shared, and broad-based knowledge to be developed and disseminated. Government funding serves as a third source of research dollars.

From its headquarters in Denver, Colorado, the Foundation's staff directs and supports the efforts of more than 800 volunteers who serve on the board of trustees and various committees. These volunteers represent many facets of the water industry, and contribute their expertise to select and monitor research studies that benefit the entire drinking water community.

The results of research are disseminated through a number of channels, including reports, the Web site, conferences, and periodicals.

For subscribers, the Foundation serves as a cooperative program in which water suppliers unite to pool their resources. By applying Foundation research findings, these water suppliers can save substantial costs and stay on the leading edge of drinking water science and technology. Since its inception, AwwaRF has supplied the water community with more than $300 million in applied research.

More information about the Foundation and how to become a subscriber is available on the Web at **www.awwarf.org**.

Hexavalent Chromium Removal Using Anion Exchange and Reduction With Coagulation and Filtration

Prepared by:
Michael J. McGuire, **Nicole K. Blute**, and **Gang Qin**
McGuire Malcolm Pirnie
1821 Wilshire Boulevard, Suite 302, Santa Monica, CA 90403

and

Peter Kavounas, **Donald Froelich**, and **Leighton Fong**
City of Glendale Water and Power
141 North Glendale Avenue, 4th Floor, Glendale, CA 91206

Jointly sponsored by:
Awwa Research Foundation
6666 West Quincy Avenue, Denver, CO 80235-3098

and

City of Glendale Water and Power
141 N. Glendale Avenue, Glendale, CA 91206

Additional support provided by the U.S. Environmental Protection Agency, Region 9

Published by:

Awwa Research Foundation

American Water Works Association

IWA Publishing

DISCLAIMER

This study was jointly funded by the Awwa Research Foundation (AwwaRF) and the City of Glendale, Calif., Water and Power (Glendale), with additional support provided by the U.S. Environmental Protection Agency (USEPA), Region 9. AwwaRF, Glendale, and USEPA assume no responsibility for the content of the research study reported in this publication or for the opinions or statements of fact expressed in the report. The mention of trade names for commercial products does not represent or imply the approval or endorsement of AwwaRF, Glendale, or the USEPA. This report is presented solely for informational purposes.

CONTENTS

TABLES

FIGURES

FOREWORD

The Awwa Research Foundation (AwwaRF) is a nonprofit corporation that is dedicated to the implementation of a research effort to help utilities respond to regulatory requirements and traditional high-priority concerns of the industry. The research agenda is developed through a process of consultation with subscribers and drinking water professionals. Under the umbrella of a Strategic Research Plan, the Research Advisory Council prioritizes the suggested projects based upon current and future needs, applicability, and past work; the recommendations are forwarded to the Board of Trustees for final selection. The foundation also sponsors research projects through an unsolicited proposal process; the Collaborative Research, Research Applications, and Tailored Collaboration programs; and various joint research efforts with organizations such as the U.S. Environmental Protection Agency, the U.S. Bureau of Reclamation, and the Association of California Water Agencies.

This publication is a result of one of these sponsored studies, and it is hoped that its findings will be applied in communities throughout the world. The following report serves not only as a means of communicating the results of the water industry's centralized research program but also as a tool to enlist the further support of the nonmember utilities and individuals.

Projects are managed closely from their inception to the final report by the foundation's staff and large cadre of volunteers who willingly contribute their time and expertise. The foundation serves a planning and management function and awards contracts to other institutions such as water utilities, universities, and engineering firms. The funding for this research effort comes primarily from the Subscription Program, through which water utilities subscribe to the research program and make an annual payment proportionate to the volume of water they deliver and consultants and manufacturers subscribe based on their annual billings. The program offers a cost-effective and fair method for funding research in the public interest.

A broad spectrum of water supply issues is addressed by the foundation's research agenda: resources, treatment and operations, distribution and storage, water quality and analysis, toxicology, economics, and management. The ultimate purpose of the coordinated effort is to assist water suppliers to provide the highest possible quality of water economically and reliably. The true benefits are realized when the results are implemented at the utility level. The foundation's trustees are pleased to offer this publication as a contribution toward that end.

David E. Rager
Chair, Board of Trustees
Awwa Research Foundation

Robert C. Renner, P.E.
Executive Director
Awwa Research Foundation

ACKNOWLEDGMENTS

The authors of this report are indebted to the following agencies, utilities, and individuals for their contributions to this project:

City of Glendale Water and Power, Glendale, Calif.
Utah State University, Logan, Utah, Laurie McNeill
University of California at Los Angeles, Los Angeles, Calif., Mel Suffet
University of Colorado at Boulder, Boulder, Colo., Gary Amy (now at UNESCO)
Lehigh University, Bethlehem, Pa., Arup SenGupta
Wellesley College, Wellesley, Mass., Daniel Brabander and James Besancon

In addition, the authors thank the Project Advisory Committee (PAC) – including Pankaj Parekh, Director of Water Quality Compliance of the Los Angeles Department of Water and Power, Los Angeles, Calif.; Sun Liang, Engineering Compliance Manager of the Water Quality Section of the Metropolitan Water District of Southern California, Los Angeles, Calif.; Richard Sakaji, Manager of Planning and Analysis for Water Quality of the East Bay Municipal Utility District, Oakland, Calif. (formerly with the California Department of Health Services in Sacramento, Calif.); and Bruce Macler, Project Officer, U.S. Environmental Protection Agency.

Portions of this work were performed at GeoSoilEnviroCARS (Sector 13), Advanced Photon Source (APS), Argonne National Laboratory. GeoSoilEnviroCARS is supported by the National Science Foundation - Earth Sciences (EAR-0622171), Department of Energy - Geosciences (DE-FG02-94ER14466) and the State of Illinois. Use of the Advanced Photon Source was supported by the U.S. Department of Energy, Office of Science, Office of Basic Energy Sciences, under Contract No. DE-AC02-06CH11357.

The authors also wish to thank AwwaRF project manager Traci Case for her assistance.

EXECUTIVE SUMMARY

Public concern about hexavalent chromium [Cr(VI)] in drinking water supplies and potential adverse health effects, coupled with California's intention to set a Cr(VI) specific drinking water standard, have prompted the investigation of cost-effective treatment technologies for Cr(VI) removal to well below the current federal and California maximum contaminant level (MCL) for total chromium (i.e., the federal MCL of 100 μg/L and the California MCL of 50 μg/L).

Following the discovery of high Cr(VI) concentrations in several groundwater extraction wells and plumes heading toward production wells, the City of Glendale, Calif. (Glendale), in partnership with the Cities of Los Angeles, Burbank, and San Fernando, devoted significant resources to identify effective Cr(VI) removal technologies from drinking water. Glendale initiated a four-phase research program, which includes: (1) the Phase I Bench-Scale Study that improved the understanding of fundamental chromium chemistry and screened promising treatment technologies, (2) the Phase II Pilot-Scale Study that evaluated Cr(VI) treatment technologies under field conditions, (3) the Phase III Bridge Project and Demonstration-Scale Study to further test a promising technology, to construct a demonstration-scale treatment facility, and to finalize the technology and cost evaluation, and (4) the Phase IV Full-Scale Implementation. Phases I and II of the program are complete and the study results were published in reports (Brandhuber et al. 2004, MEC 2005) and peer-reviewed journal articles (Qin et al. 2005, McGuire et al. 2006).

In the Phase II Pilot-Scale Study, weak-base anion (WBA) exchange resin demonstrated an unexpectedly high Cr(VI) removal capacity that might make its use cost-effective as a disposable media. Before the WBA resin was considered for testing in the Phase III Demonstration-Scale Study, however, a study designated as the Phase III Bridge Project was conducted to further investigate the removal of Cr(VI) with WBA resins. This report presents the findings and results of the Phase III Bridge Project, which was intended to "bridge" pilot-scale and demonstration-scale testing. Key objectives of the Phase III Bridge Project included:

- Conducting treatment studies to confirm the efficiencies of WBA resins for Cr(VI) removal from Glendale groundwater
- Characterizing WBA resin residuals to elucidate Cr(VI) removal mechanism(s)
- Investigating residuals handling and disposal options
- Refining cost estimates of effective Cr(VI) treatment technologies (WBA, strong-base anion (SBA) exchange resin, and reduction/coagulation/filtration (RCF)
- Convening an expert panel to recommend treatment technologies for demonstration-scale testing

The Phase III Bridge Project began with a bench-scale isotherm evaluation to screen six promising WBA resins for Cr(VI) removal from spiked Glendale groundwater at two pH values (5.9 and 6.4). Table ES.1 lists the names and characteristics of the six resins tested. SIR-700 and Duolite A7 resins outperformed the other four WBA resins for Cr(VI) removal at both pH conditions and were selected for subsequent flow-through pilot testing. Of note in isotherm testing was that both resins required more than 64 days to reach equilibrium with Cr(VI) in solution. The high Cr(VI) capacities of the SIR-700 and Duolite A7 resins coupled with the slow kinetics to reach

Table ES.1
Weak-base anion exchange resins evaluated in bench-scale isotherm testing

Manufacturer	Resin name	Matrix	Functional group
Rohm & Haas	Duolite A7	Phenol-formaldehyde polycondensate	Secondary amine
ResinTech	SIR-700	Epoxy polyamine	Proprietary amine
Sybron	Lewatit S4528	Styrene-divinylbenzene (macroporous)	Tertiary & quaternary amine
Purolite	A146	Styrene-divinylbenzene (macroporous)	Tertiary amine
Purolite	A830	Polyacrylic-divinylbenzene (macroporous)	Complex amine
Dow	Monosphere 66	Styrene-divinylbenzene (macroporous)	Tertiary amine

equilibrium indicated that a mechanism other than ion exchange might be involved in Cr(VI) removal by the WBA resins.

Bench-scale isotherm testing was followed up by short-term mini-column (0.5-inch* diameter) testing and longer-term pilot-scale column (2.5-inch diameter) testing. The selection of an appropriate pH for pilot-scale testing was determined using mini-columns for ten days. Mini-column testing results for pH values ranging from 5.6 to 7.2 showed different breakthrough characteristics for the two resins tested (SIR-700 and Duolite A7). However, both resins favored low pH for Cr(VI) removal to achieve a treatment goal of less than 5 µg/L. A pH of 6.0 was selected as the reduced pH at which to test Cr(VI) removal capacities in the longer-term pilot tests.

SIR-700 and Duolite A7 resins were tested in pilot-scale columns at a pH of 6.0 and ambient pH of 6.8, the latter of which was examined to evaluate the potential for reduced acid addition but more frequent resin replacement. Figure ES.1 shows the Cr(VI) breakthrough curves of both resins, highlighting the importance of a reduced pH for Cr(VI) removal. At pH 6.0, the Duolite A7 resin treated approximately 45,000 bed volumes (BV) of water before 5 µg/L Cr(VI) (and total Cr) effluent concentrations were observed. The consistent removal performance over an extended period indicated that the Duolite A7 resin could be used as an effective single-pass resin for Cr(VI) removal in Glendale groundwater.

Although the SIR-700 resin treating pH 6.0 water showed an early 5 µg/L breakthrough point at 2,200 BV, the Cr(VI) removal performance improved during the testing period such that effluent total Cr and Cr(VI) concentrations were less than 5 µg/L at the end of the pilot-scale testing (i.e., after treating approximately 113,000 BV of water). Improved Cr(VI) removal through the testing is not typical of ion exchange breakthrough curves, suggesting another removal mechanism may have contributed significantly to Cr(VI) removal or the resin required additional conditioning prior to use.

---

* Note: Information regarding SI units and U.S. customary units appear in Appendix A.

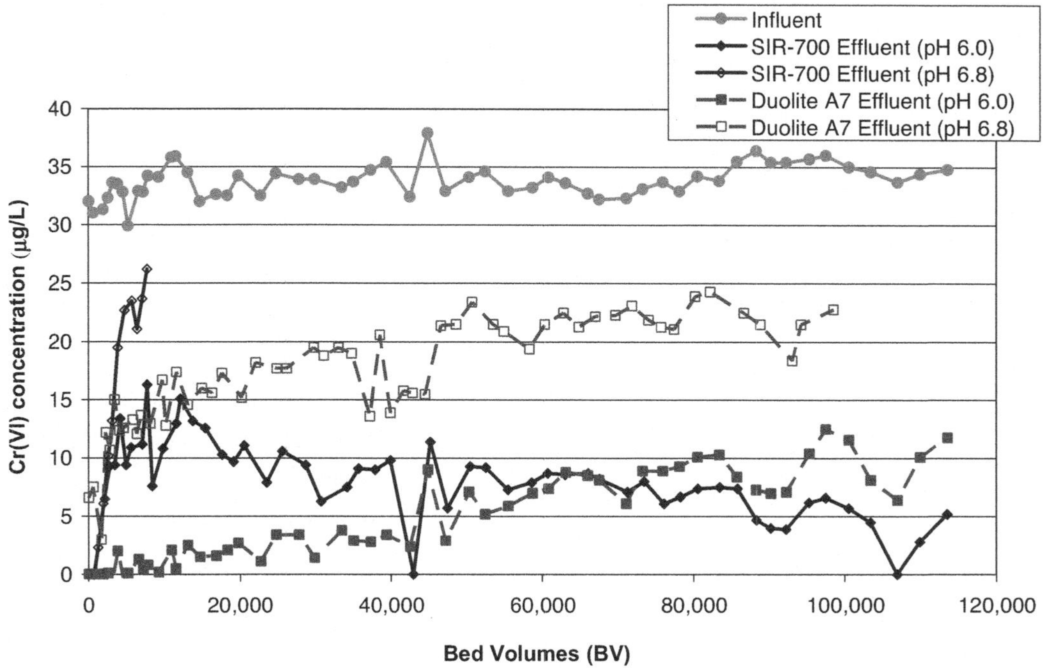

Figure ES.1 Pilot-scale column breakthrough curves of Cr(VI) at pH 6.0 and ambient pH (6.8) for Duolite A7 and SIR-700 resins

Once pilot testing was complete, spent resins were removed and tested for hazardous waste characteristics. The spent SIR-700 and Duolite A7 resins operated at pH 6.0 passed the federal toxicity characteristic leaching procedure (TCLP) but failed the California Waste Extraction Test (WET) and thus would be characterized as probable hazardous wastes for disposal in California. The spent Duolite A7 resins also had total uranium concentrations exceeding 500 µg/g (i.e., the trigger for low-level radioactive waste designation) after treating approximately 113,000 BV of water, indicating that the operating life of the Duolite A7 resin may need to be limited to avoid generating a low-level radioactive waste.

Mechanisms for Cr(VI) removal with WBA resins were also investigated in the Phase III Bridge Project. Chromium speciation on spent resins was assessed using x-ray absorption near-edge structure (XANES) spectroscopy. Figure ES.2 shows the Cr XANES spectra of resin residuals samples and known trivalent chromium [Cr(III)] and Cr(VI) reference compounds. The Cr spectral overlap of the spent resins and the trivalent chromium reference compound [Cr(III) acetate] and the lack of any pre-edge absorption peak indicated that Cr(III) was the dominant species retained on both resins, comprising more than 95% of the total chromium present. The XANES analysis provided direct evidence that the reduction of Cr(VI) to Cr(III) was an important part of the Cr(VI) removal mechanism by the SIR-700 and Duolite A7 WBA resins.

In addition to the technical evaluations, cost estimates of WBA resin application at the demonstration-scale were developed and compared with the other two promising Cr(VI) removal technologies: SBA and RCF. Detailed annualized cost estimates of various technologies are shown in Table ES.2. Although the annualized cost estimates were lowest for the SBA option at

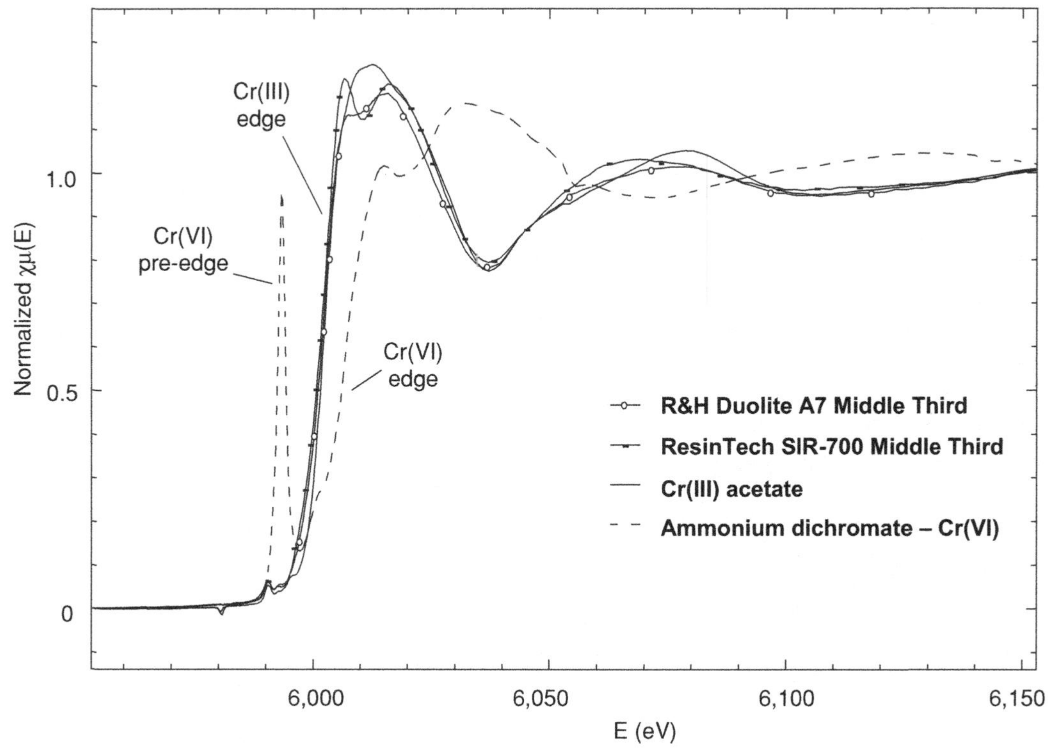

Figure ES.2 Cr XANES spectra of resin samples and selected Cr reference compounds

Table ES.2
Annualized treatment cost estimates for Cr(VI) removal

Technology	Flow (gpm)	Annualized costs ($/AF)		
		Capital	O&M	Total
WBA	500 (Retrofit)	100	340	440
	500	170	350	520
	1,000	120	340	460
SBA	500	190	170	360
	1,000	110	130	240
RCF	500	280	190	470
	1,000	180	120	300

both potential flow rates (500 gpm and 1,000 gpm), the uncertain future of brine disposal could make the SBA process cost-prohibitive. At 500 gpm, a retrofit WBA system (i.e., converting two existing GAC contactors to ion exchange vessels at the Glendale GS-3 well site) was determined to be a cost-effective Cr(VI) treatment technology implementation for Glendale. For a 1,000 gpm system, the RCF process is more cost-effective than WBA resin due largely to lower operations and maintenance (O&M) costs.

After reviewing the technical and cost information from the Phase I, Phase II, and Phase III Bridge Project, an expert panel concluded that the RCF system should be tested in the Phase III Demonstration-Scale Study. Further investigation of the Cr(VI) removal mechanism by the WBA resins was recommended; consequently, additional bench-scale studies will be conducted prior to the demonstration-scale study. Demonstration-scale testing of SBA resin was not advised by the panel.

Based on the results from the Phase III Bridge Project, limited available funding for capital costs, and the goal of achieving the most reduction in the water supply's Cr(VI) levels, the City of Glendale is planning to implement a 500-gpm Phase III Demonstration-Scale WBA system at the Glendale GS-3 well site by retrofitting existing vessels. Although the RCF process was recommended by the expert panel, the relatively high capital costs associated with this technology might only allow the City of Glendale to build a smaller-scale RCF treatment system, which is planned in conjunction with the WBA retrofit. An RCF system would be installed for well treatment at a location adjacent to the Glendale Water Treatment Plant (GWTP).

CHAPTER 1
INTRODUCTION

BACKGROUND

The presence of volatile organic compounds (VOCs) in the San Fernando Valley groundwater basin has resulted in a combined clean-up effort between the USEPA, the State, local agencies, and potentially responsible parties. Four Superfund sites were identified and included in the Superfund National Priorities List and were later divided into six operable units: North Hollywood, Burbank, Headworks, Glendale-North (GN), Glendale-South (GS), and Pollock. The Glendale Water Treatment Plant (GWTP) was designed to treat VOCs from 8 wells at the Glendale-North and Glendale-South Operable Units and ultimately serve the treated water to customers in Glendale. The unit processes employed in the treatment plant include packed tower air stripping with vapor phase granular activated carbon (GAC) treatment followed by liquid-phase GAC. In addition to VOC contamination, low-level chromium concentrations were also measured in the groundwater samples. However, during treatment facility design, exposure to VOCs in the groundwater constituted the primary chronic human health risk and the treatment plants were not designed to remove chromium.

In 2000, as the City of Glendale Calif. was about to begin delivery of treated water from the GWTP, the release of the movie *Erin Brockovich* generated public concern about the presence of hexavalent chromium [Cr(VI)] in drinking water even though concentrations were far less than the federal maximum contaminant level (MCL, 100 µg/L for total chromium) and the California MCL (50 µg/L for total chromium). Glendale has managed to deliver finished water with Cr(VI) levels at approximately 5 µg/L by blending the GWTP effluent with imported water from the Metropolitan Water District of Southern California (MWDSC) and, under short term approval from the USEPA, by pumping more water from the low Cr(VI) wells and reducing pumping from the high Cr(VI) wells. However, Cr(VI) concentrations are projected to increase in some of the supply wells (e.g. historical data for GN-2, GN-3, and GS-3 wells shown in Figure 1.1) as contaminant plumes advance toward the wells. In addition, the state of California is mandated to set a Cr(VI) MCL, which is expected to be lower than the total chromium MCL. Until recently, very little information was available on the capabilities of various treatment technologies to remove Cr(VI) to low µg/L levels. In response to all the uncertainties and concerns, the City of Glendale, in partnership with the Cities of Los Angeles, Burbank, and San Fernando, initiated a four-phase program to develop a full-scale Cr(VI) treatment system capable of yielding low chromium levels in the drinking water supply.

FOUR-PHASE TREATMENT PROGRAM

The four-phase program developed by the City of Glendale is summarized below.

Phase I – Bench-Scale Study

A bench-scale study was initiated by the City of Glendale in partnership with the Cities of Los Angeles, Burbank and San Fernando, the Awwa Research Foundation (AwwaRF) and the National Water Research Institute (NWRI). The focus of this study was to improve the

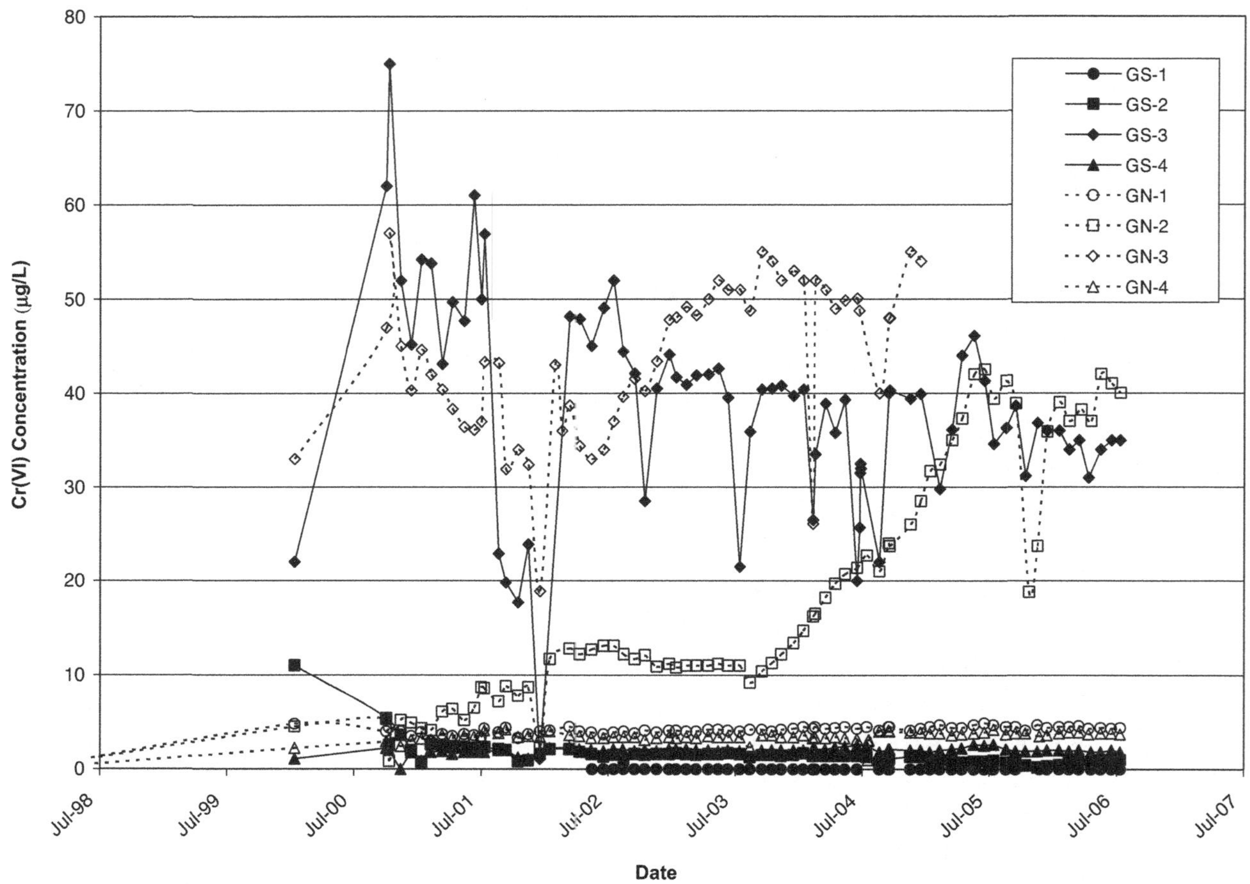

Figure 1.1 Cr(VI) concentrations in Glendale extraction wells

understanding of fundamental chromium chemistry and to screen a large number of promising treatment technologies for their ability to treat and remove Cr(VI) to very low levels.

Based on the bench-scale study, several technologies were identified as promising for utility application. Further study of the technologies listed below was recommended to characterize their performance under flow-through pilot testing conditions:

- Anion exchange (both as fixed-bed and dispersed contactor applications; strong-base and weak-base anion exchange resins)
- Sulfur modified iron sorption media
- Coagulation and precipitation of reduced Cr(III)

While the removal of Cr(VI) using nanofiltration and reverse osmosis membranes was shown to be technically effective, membrane treatment was not recommended due to the large loss of water associated with this technology.

The Phase I Bench-Scale Study is complete and a final report was published by AwwaRF (Brandhuber et al. 2004).

Phase II – Pilot-Scale Study

The Phase II Pilot-Scale Study was funded by Congressional appropriation and administered as a grant by the USEPA. The objective of the Phase II study was to demonstrate two categories of treatment technologies at pilot-scale: Cr(VI) treatment technologies shown to be effective based on Phase I bench-scale testing (Phase A) and emerging treatment technologies or residuals minimization strategies for effective technologies (Phase B). The study was performed by the City of Glendale and McGuire Environmental Consultants, Inc. (MEC) with the assistance of Utah State University (USU), University of California at Los Angeles (UCLA) and the University of Colorado at Boulder (CU).

Phase A identified both effective and ineffective treatment technologies for Cr(VI) removal. A pilot testing treatment goal of 5 μg/L was established for a number of reasons: to test technology performance with respect to Glendale's low treatment targets, to represent 95% removal when treating a 100 μg/L Cr(VI) influent water, and to enable rigorous evaluation of the goal without being too close to the detection limits for Cr(VI) and total Cr. Strong-base anion (SBA) exchange resins provided by several vendors showed a range of capacities, but demonstrated that SBA resin could achieve the treatment goal of 5 μg/L. A reactor-based SBA resin system (MIEX, Orica WaterCare) was inconsistent – i.e., able to remove 95% of the Cr(VI) about half of the time during the pilot testing. By contrast with SBA resins, weak-base anion (WBA) exchange resins showed a capacity approximately twenty times greater. The high capacity of the WBA resin was an unexpected finding from Phase A. Additional work was recommended (the Phase III Bridge Project) to evaluate WBA resins using a reliable pH adjustment system, as this component was problematic in the Phase A study.

Several technologies were ineffective in achieving the Cr(VI) treatment goal or impractical, including a reduction-filtration approach, iron-impregnated GAC, and zeolite media. One technology tested by a participating vendor attempted to reduce Cr(VI) to Cr(III) with high doses of sodium metabisulfite (and later sodium sulfite), then add chlorine to the water to obtain a slightly oxidized environment before Cr(III) was filtered by a proprietary filter bed. The high concentrations of chemicals required, in addition to ineffective total chromium removal by the filter bed, made this technology unsuccessful. Iron-impregnated GAC and zeolite media proved impractical because the quantity of water treated before exhaustion was significantly lower than for SBA or WBA resins.

Phase B tested both the reduction/coagulation/filtration (RCF) process (which was in design for the Hinkley treatment system) and regeneration of SBA resins using recycled brine to determine if hazardous brine could be minimized. Pilot testing of the RCF process was very effective, indicating that this technology holds significant promise for Cr(VI) treatment. Phase B testing also demonstrated that SBA resins could be regenerated multiple times using recycled brine.

Finally, the Phase II Pilot-Scale Study developed unit costs for each of the successful technologies, including SBA resin, WBA resin, and RCF.

The Phase II Pilot-Scale Study is complete and a final report was submitted to the City of Glendale (MEC 2005). Selected results were also published in peer-reviewed scientific journals (Qin et al. 2005, McGuire et al. 2006).

Phase III – Bridge Project and Demonstration-Scale Study

The Phase III Demonstration-Scale Study will finalize the treatment evaluation, residuals assessment, and cost model development by implementing one or more effective technologies at demonstration-scale for the treatment of Cr(VI) containing well(s) at either the GWTP or a Glendale well site. The initial phase of the Phase III effort was designated as the Phase III Bridge Project, which included additional studies to finalize testing of weak-base anion exchange resins for Cr(VI) treatment, refinement of treatment technology cost estimates based on Phase III Bridge Project results, and assembly of an expert panel to recommend the one or more treatment processes for demonstration-scale installation. Funding of the Phase III Bridge Project was provided by an AwwaRF Tailored Collaboration (TC) grant and a FY 2003 Congressional appropriation. Results of the Phase III Bridge Project are presented in this report.

Phase IV – Full-Scale Implementation

If necessary to reduce Cr(VI) concentrations further, the final phase of the four-phase testing program will implement full-scale treatment for Cr(VI) removal at the City of Glendale based upon the results of the Phase III Demonstration-Scale Study.

PROJECT OBJECTIVES

As a result of the Phase II Pilot-Scale Study, three effective Cr(VI) treatment technologies were identified: regenerable SBA resin, WBA resin, and RCF. Among the technologies, weak-base anion exchange was distinctive in its high capacity for Cr(VI) removal and represented an innovative technology application. Although the mechanism of Cr(VI) removal was yet unknown, WBA resin demonstrated an extraordinary capacity for Cr(VI) that was far superior to other anion exchange resins or adsorptive media. The Phase III Bridge Project was designed to further characterize the efficiency of these weak-base anion exchange resins and to compare the economics of their one-time use with the use of regenerable SBA resins (including brine disposal costs). The research results would then be used to finalize the selection of the technologies for demonstration-scale installation.

The overall objectives of the Bridge Project included the following:

- Conducting treatment studies to confirm the efficiencies of WBA resins for Cr(VI) removal from Glendale groundwater
- Characterizing WBA resin residuals and investigating their handling and disposal
- Refining cost estimates of the three effective Cr(VI) treatment technologies for demonstration-scale installation
- Convening an expert panel to recommend the most cost-effective treatment method(s) for use in demonstration-scale testing

REPORT ORGANIZATION

The project was divided into several major tasks, and the results of each task are presented as separate chapters in this report. The organization of the report is listed below:

- Bench-scale isotherm testing (Chapter 2)
- Mini-column evaluation (Chapter 3)
- Pilot-scale column evaluation (Chapter 4)
- Residuals analyses (Chapter 5)
- Cost estimates (Chapter 6)
- Expert panel discussions (Chapter 7)
- Summary and conclusions (Chapter 8)

CHAPTER 2
BENCH-SCALE TESTING

Ion exchange is generally a physical/chemical process in which an ion with high affinity for the ion exchange resin replaces a previously bound, lower affinity ion (Jacobs and Testa 2005). Anion exchange resins have been used in industrial applications to treat Cr(VI) waste water. In conventional anion exchange processes, Cr(VI) [in the form of chromate (CrO_4^{2-}), bichromate ($HCrO_4^-$), or dichromate ($Cr_2O_7^{2-}$)] replaces Cl^- or OH^- ions previously bound to the resins. For convenience in the discussion, the different chromate species will be presented as Cr(VI). The individual anion chemical formulas will be used for individual species whenever necessary.

Anion exchange is a proven industrial treatment technology for influent Cr(VI) concentrations in the microgram- to milligram-per-liter range (Richardson, Stobbe, and Bernstein 1968; Jakobsen and Laska 1977; Patterson 1985; Sengupta, Clifford, and Subramonian 1986; Sengupta and Clifford 1986). In the past decade, the potential use of anion exchange resins to remove Cr(VI) to low microgram-per-liter levels from drinking water supplies has received more attention (Bahowick, Dobie, and Kumamoto 1996; Clifford 1999; Höll et al. 2002; Brandhuber et al. 2004; McGuire et al. 2006). These studies demonstrated that anion exchange was capable of treating Cr(VI) to very low ppb levels.

Two primary categories of anion exchange resins – strong-base and weak-base – can be used for anionic contaminant removal from water. For Cr(VI) treatment, SBA resins typically have quaternary amine functional groups and can be regenerated to some degree with a salt solution. SBA resins are commonly used in drinking water treatment for removal of other anions besides Cr(VI). By comparison, WBA resins typically have tertiary or secondary amine functional groups and are proposed as disposable media for Cr(VI) removal due to challenges in regenerating (i.e., requirements of acids and/or bases rather than salt to remove contaminants). Thus far, WBA resins have not been used for full-scale Cr(VI) removal from drinking water.

A surprising finding from the Phase II Pilot-Scale Study conducted by McGuire et al. (2006) was that a WBA resin (Duolite A7, provided by Siemens in partnership with Rohm & Haas) demonstrated a Cr(VI) removal capacity approximately 20 times higher than that of the conventional SBA resins (Figure 2.1). A similar Cr(VI) removal performance with the same WBA resin was also reported by Höll et al. (2002).

Since Cr(VI) removal using WBA resins held significant potential as a cost-effective disposable media but was still a new application in drinking water treatment, more research was needed before launching a demonstration-scale test. The Phase III Bridge Project was designed to investigate the WBA resins and included bench-scale testing, mini-column testing, and pilot-scale testing. The first task of the Phase III Bridge Project was to screen a number of WBA resins using isotherm testing.

EXPERIMENTAL METHODOLOGY

Six promising WBA resins from different resin manufacturers were selected for bench-scale isotherm testing based on past Cr(VI) removal performances, commercial availability, and vendor recommendations. The six resins are listed and characterized in Table 2.1. Note that the mention of resin names does not imply endorsement by the authors. The purpose of bench-scale

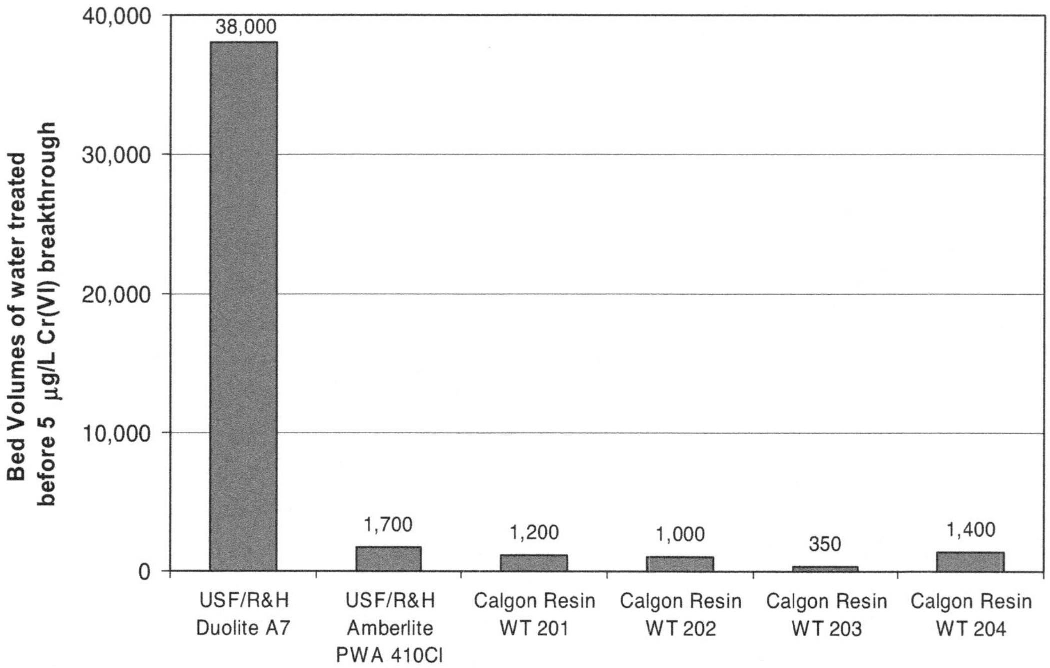

Notes:
1. Results correspond to treatment of an influent Cr(VI) concentration of 100 µg/L.
2. Except WBA Duolite A7, all other resins were SBA resins.

Figure 2.1 Cr(VI) removal performance of ion exchange resins in Phase II pilot-scale studies

Table 2.1
Properties of the WBA resins evaluated in bench-scale isotherm testing

Manufacturer	Resin name	Matrix	Functional group
Rohm & Haas	Duolite A7	Phenol-formaldehyde polycondensate	Secondary amine
ResinTech	SIR-700	Epoxy polyamine	Proprietary amine
Sybron	Lewatit S4528	Styrene-divinylbenzene (macroporous)	Tertiary & quaternary amine
Purolite	A146	Styrene-divinylbenzene (macroporous)	Tertiary amine
Purolite	A830	Polyacrylic-divinylbenzene (macroporous)	Complex amine
Dow	Monosphere 66	Styrene-divinylbenzene (macroporous)	Tertiary amine

Table 2.2
General water quality parameters of the Glendale GS-3 well water

Constituents	Typical concentration
Alkalinity	200 mg/L as $CaCO_3$
Arsenic (total)	< 2 μg/L
Chromium (total)	35–40 μg/L
Chromium (hexavalent)	35–40 μg/L
Conductivity	850 μS/cm
Copper	20 μg/L
Hardness	350 mg/L as $CaCO_3$
Iron (total)	< 6 μg/L
Manganese	< 20 μg/L
Nitrate	7 mg/L as NO_3
pH	6.8
Phosphate	0.3 mg/L as PO_4
Silicate	33 mg/L as SiO_2
Sulfate	100 mg/L as SO_4
Uranium*	1.4 pCi/L
Vanadium	7 μg/L

*Uranium concentrations were only measured twice: in 2000 (1.48 pCi/L) and 2001 (1.39 pCi/L). Other constituents are typically measured monthly.

testing was to narrow the list of 6 resins down to the 2 with the highest capacity for subsequent testing in flow-through column testing.

Groundwater from Glendale's South Operable Unit well GS-3 was used in the bench-scale testing and in the subsequent flow-through column testing. General water quality parameters of the GS-3 groundwater are listed in Table 2.2. For isotherm tests, GS-3 well water was spiked with sodium chromate to reach a 1 mg/L Cr(VI) concentration.

The experiments also addressed the Cr removals by WBA resins at two pH values. Phase II Pilot-Scale Study testing showed that Cr(VI) removal efficiency of the Duolite A7 WBA resin was impacted by pH (McGuire et al. 2006). Based on manufacturers' recommendations and past findings, the isotherm experiments were conducted at two initial pH conditions: 5.9 and 6.4.

Isotherm experiments were conducted in 500 mL high-density polyethylene (HDPE) Nalgene bottles. At both pH 5.9 and 6.4, four resin doses were tested: 20, 27.5, 35, and 42.5 mg of resin per 500 mL GS-3 water (corresponding to resin doses of 40, 55, 70, and 85 mg/L) based on a preliminary isotherm test using the Duolite A7 resin to identify the approximate capacities that would be observed in the isotherm tests. The doses were selected to achieve detectable Cr(VI) concentrations at equilibrium. Once the known resin doses were added to the bottles containing spiked GS-3 water, the bottles were placed on a bottle tumbler rotating at 30 revolutions per minute to ensure the resins remained in suspension. After 10, 16, 24, 31, 44, and 64 days, the bottles were temporarily removed from the tumbler for sample collection. During each sampling event, 10 mL water samples were withdrawn from the bottles and filtered through a 0.45 μm filter.

Cr(VI) concentrations in the filtered sample were measured using a Hach DR/4000 spectrophotometer with the Hach 8023 1,5-diphenylcarbazide method (detection limit of approximately 6 µg/L; Hach Company, Loveland, Colo.).

In an additional bench-scale experiment, the potential for nitrosamine leaching from fresh Duolite A7 and SIR-700 resins was investigated. For this test, 150 mL of resin was added to 3 L of GS-3 water in a glass beaker [i.e., yielding a 20:1 water-to-resin ratio consistent with past nitrosamine leaching tests by Najm and Trussell (2001)]. The resin was kept in suspension for 4 hours using a magnetic stir plate. After 4 hours, 3 L samples were filtered through a 0.45 µm filter, preserved with sodium thiosulfate, and shipped to Montgomery Watson Harza Laboratories for analysis of 7 nitrosamine species listed below using EPA Method 521:

- NDMA: N-Nitrosodimethylamine
- NDEA: N-Nitrosodiethylamine
- NDPA: N-Nitrosodi-n-propylamine
- NDBA: N-Nitrosodi-n-butylamine
- NMEA: N-Nitrosomethylethylamine
- NPIP: N-Nitrosopiperidine
- NYPR: N-Nitrosopyrrolidine

RESULTS AND DISCUSSION

Cr(VI) Removal Kinetics

One important physicochemical aspect of Cr(VI) removal by WBA resins is kinetics. Typical SBA resins rapidly exchange inorganic ions, usually on the order of minutes to reach equilibrium (Horng and Clifford 1997, Clifford 1999). In contrast, ion exchange with WBA resins can be slow due to the tight, non-swollen nature of the WBA resins and may require hours to attain equilibrium (Clifford 1999). In column operations, leakage of contaminants will be more significant with media exhibiting slower kinetics.

The kinetics of Cr(VI) removal can be demonstrated by plotting Cr(VI) concentrations remaining in the aqueous phase versus time. Figures 2.2 to 2.5 show the Cr(VI) removal kinetics for different resins at resin doses from 40 mg/L to 85 mg/L at pH values of 5.9 and 6.4. As displayed in Figures 2.2 to 2.5, Cr(VI) removal by Duolite A7, SIR-700, and A146 resins appeared to require more than 64 days to reach equilibrium. Cr(VI) removal by A830, Monosphere 66, and Lewatit S4528 resins reached equilibrium in much less time. However, the capacities of these three resins were the lowest of the six tested.

Overall, Cr(VI) removal by the WBA resins was demonstrated to be a kinetically slow process, reaching equilibrium on the order of days. This result was contradictory to the common knowledge that ion exchange is a fairly rapid reaction. A potential explanation for the slow kinetics may be that another (slower) reaction may play a role in Cr(VI) removal by the WBA resins. The Phase III Bridge Project investigation of the resin mechanisms for Cr(VI) removal is discussed in detail in the following chapters.

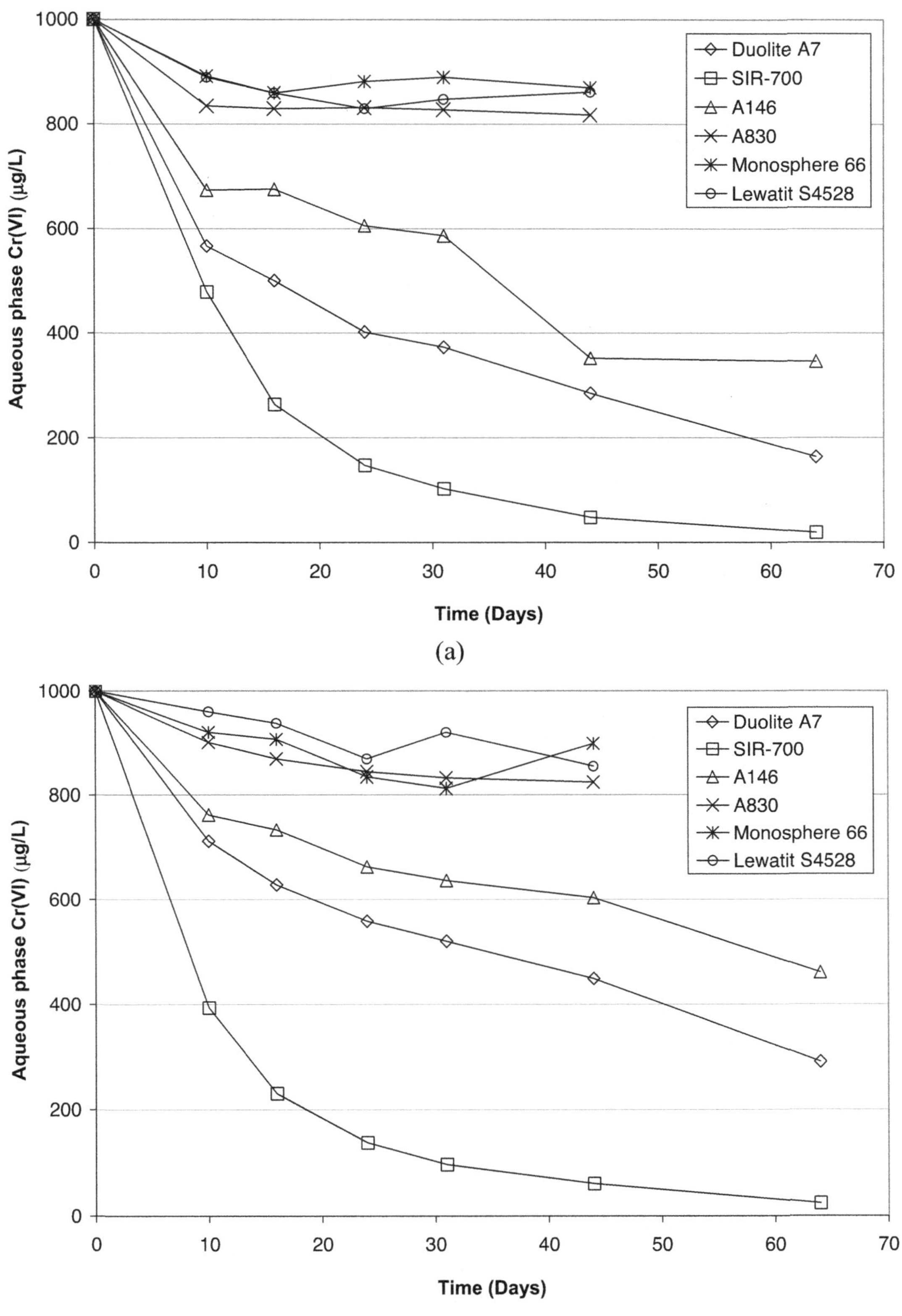

Figure 2.2 Kinetics of Cr(VI) removal by different resins at resin doses of 40 mg/L at (a) pH 5.9, and (b) pH 6.4

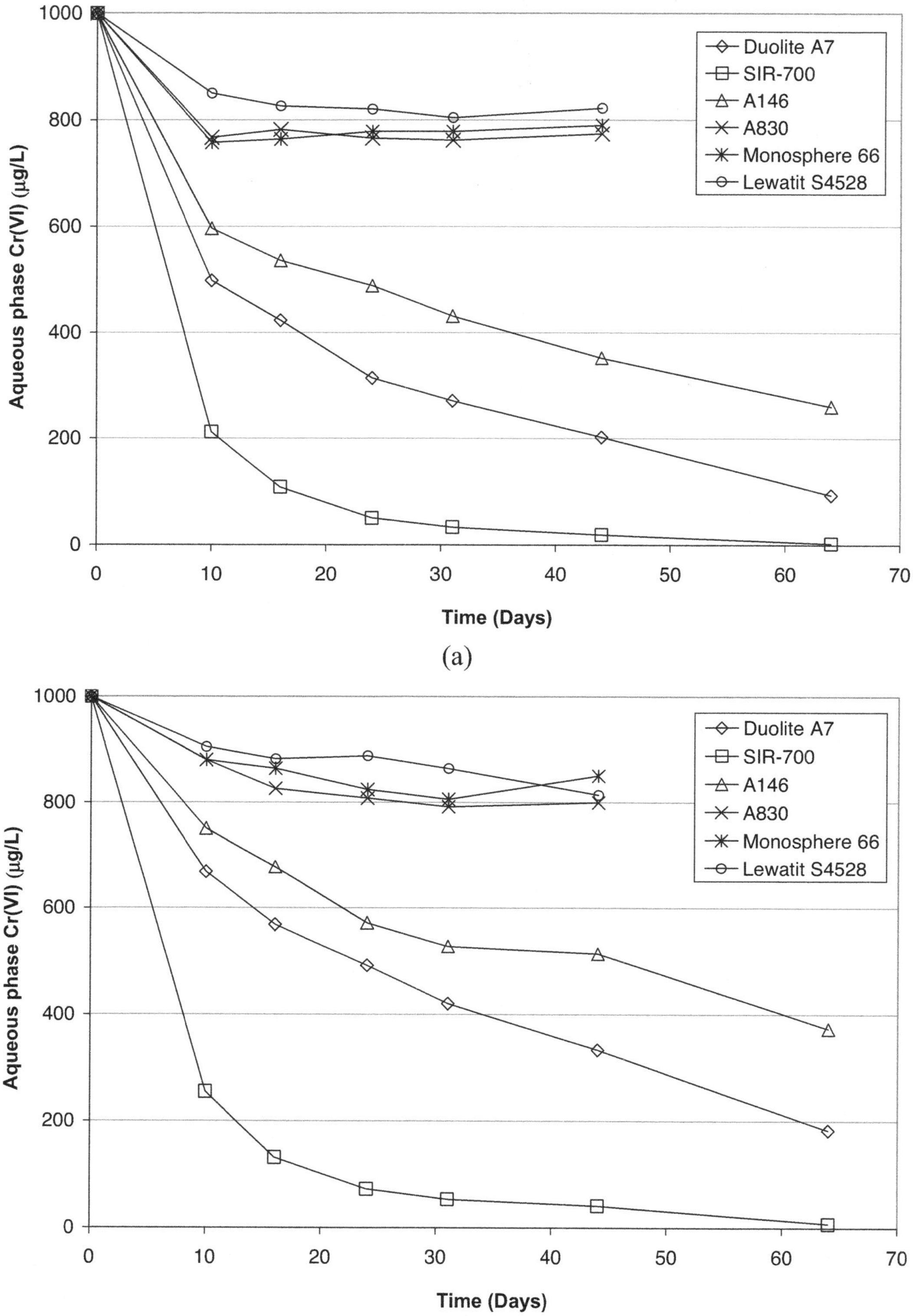

(a)

(b)

Figure 2.3 Kinetics of Cr(VI) removal by different resins at resin doses of 55 mg/L at (a) pH 5.9, and (b) pH 6.4

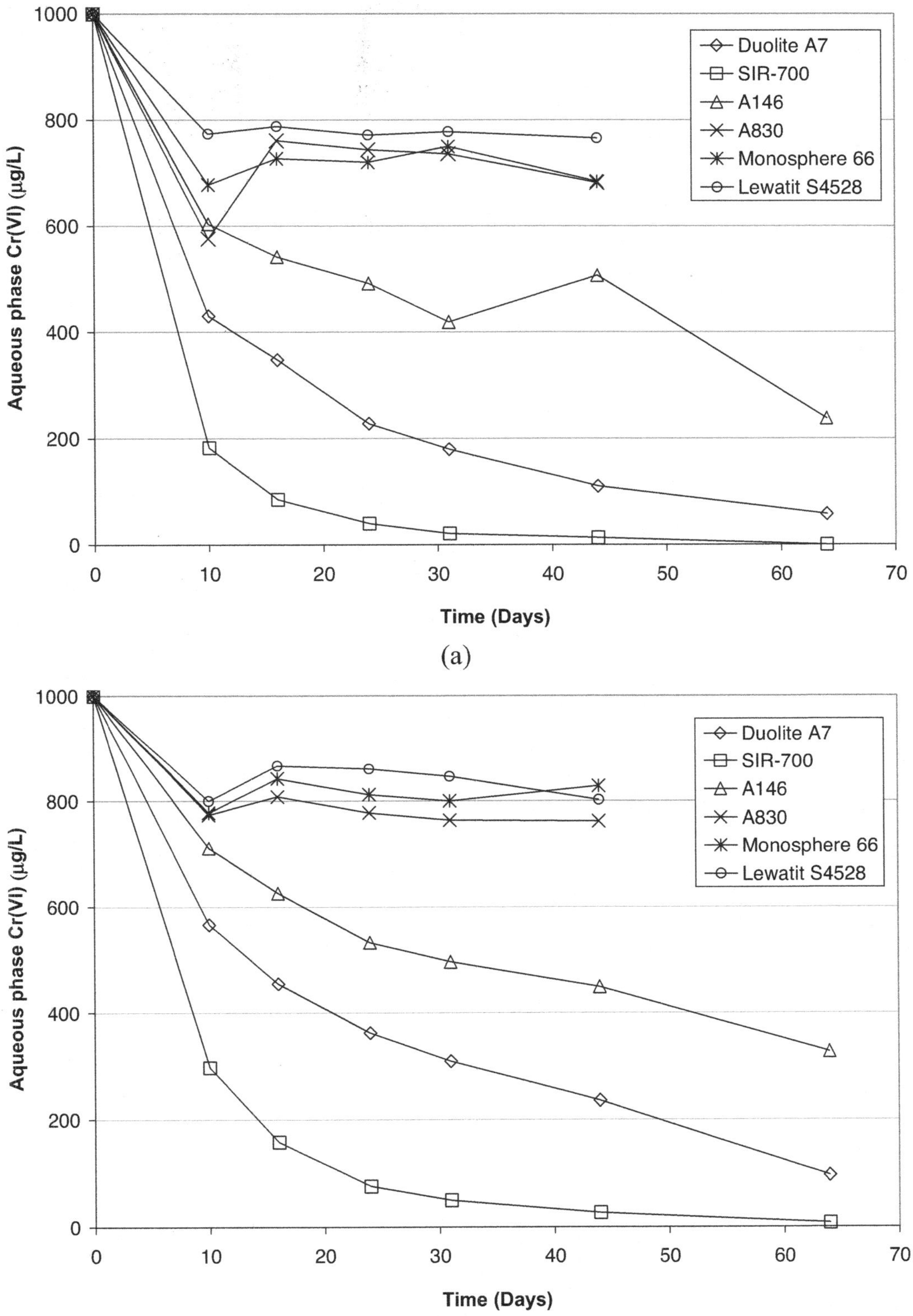

(a)

(b)

Figure 2.4 Kinetics of Cr(VI) removal by different resins at resin doses of 70 mg/L at (a) pH 5.9, and (b) pH 6.4

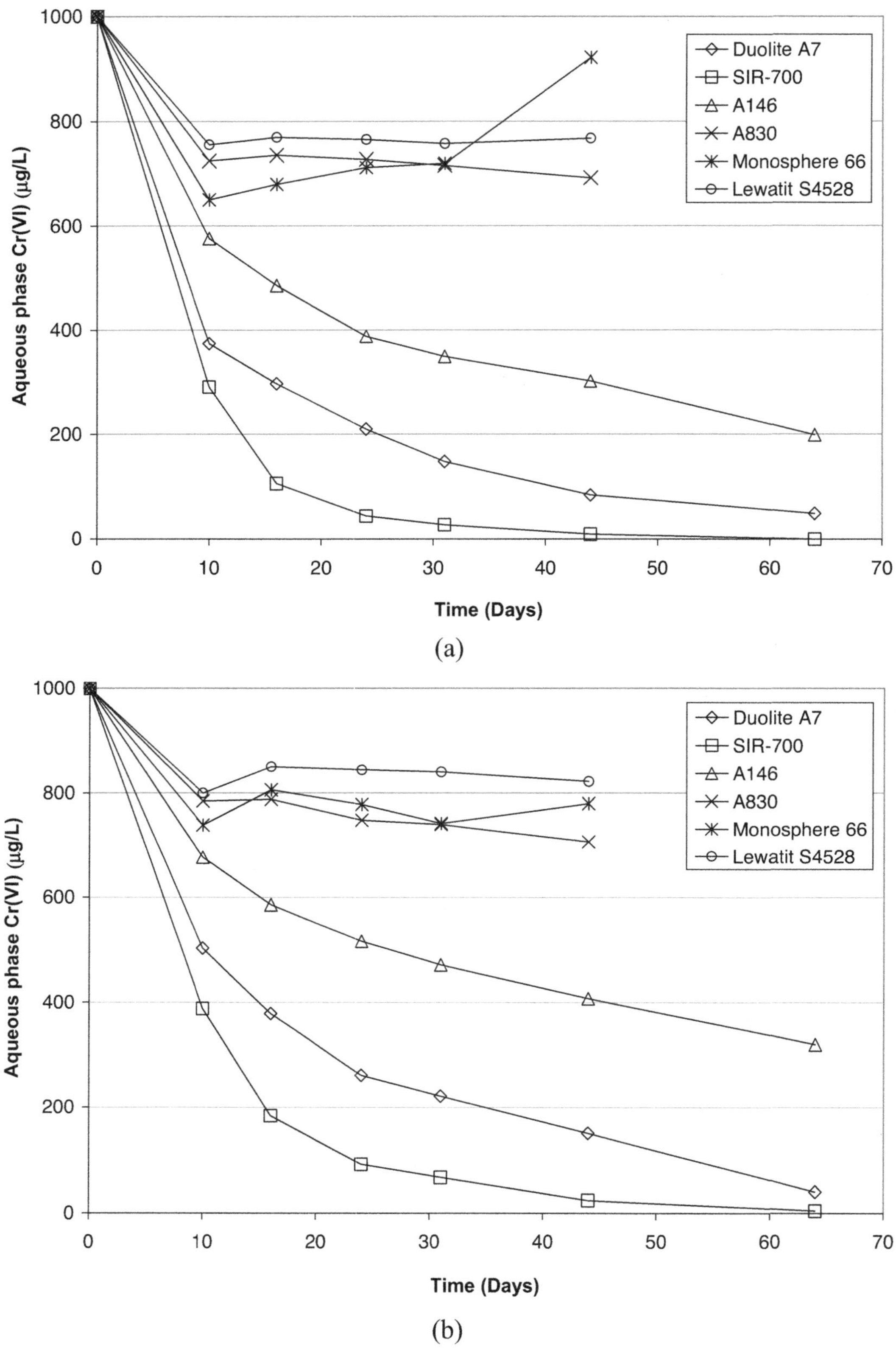

Figure 2.5 Kinetics of Cr(VI) removal by different resins at resin doses of 85 mg/L at (a) pH 5.9, and (b) pH 6.4

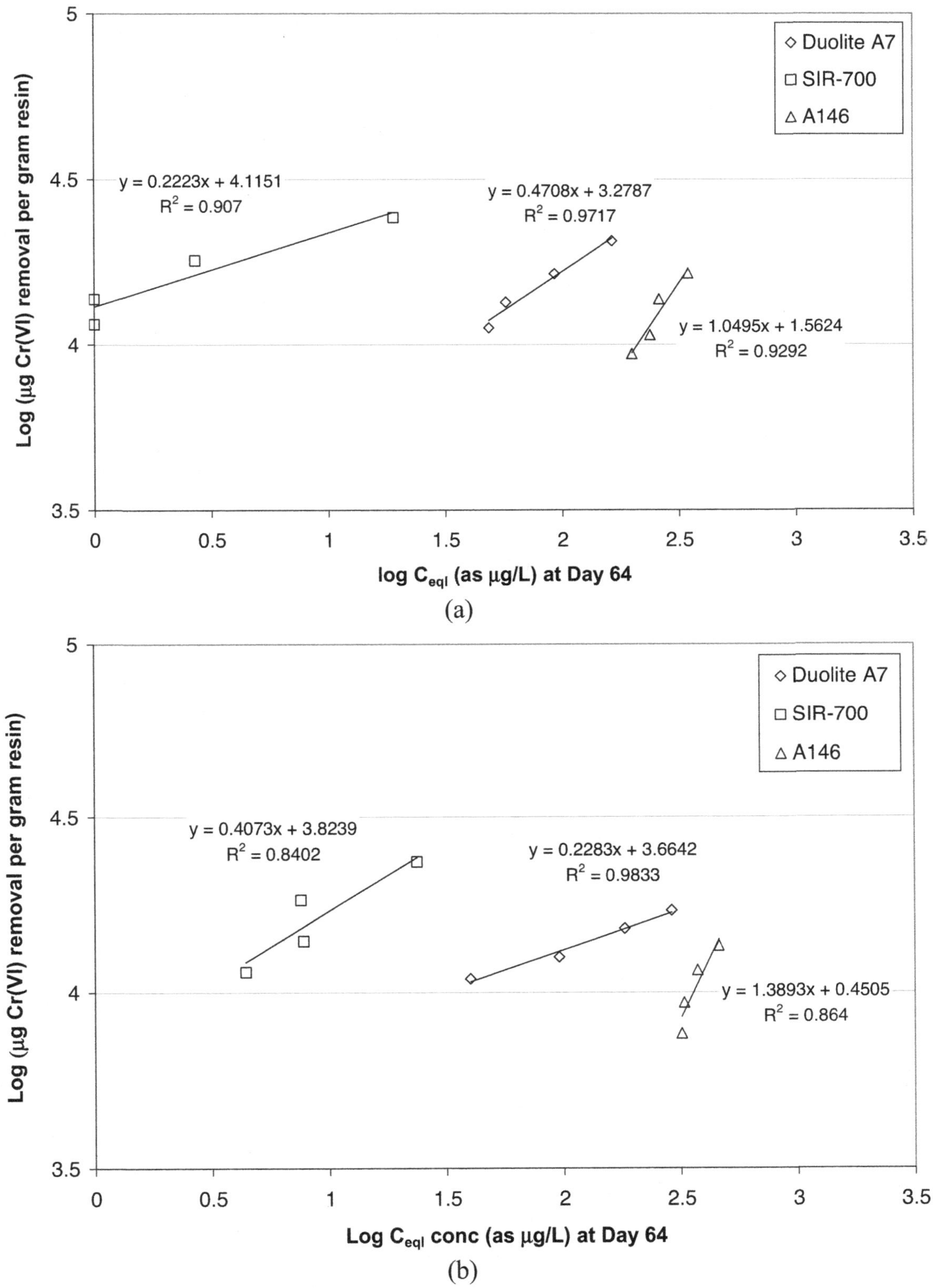

Figure 2.6 Bench testing isotherms of Cr(VI) removal by three WBA resins (Duolite A7, SIR-700, and A146) on day 64 at (a) pH 5.9 and (b) pH 6.4

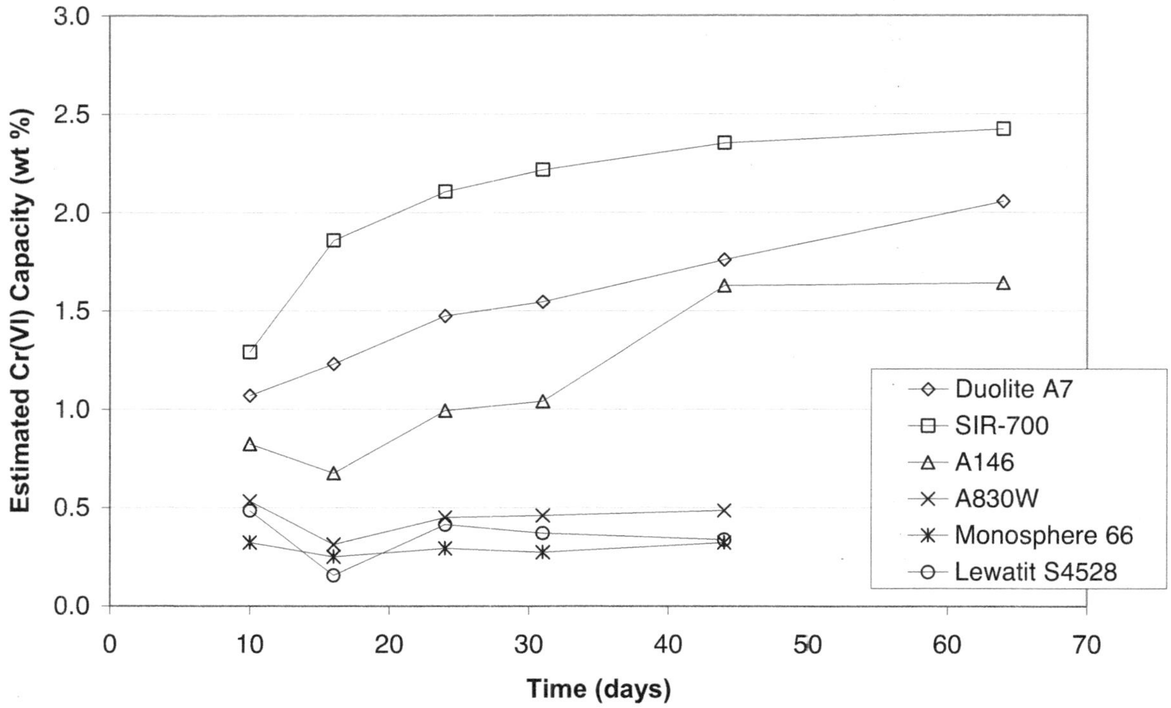

(a)

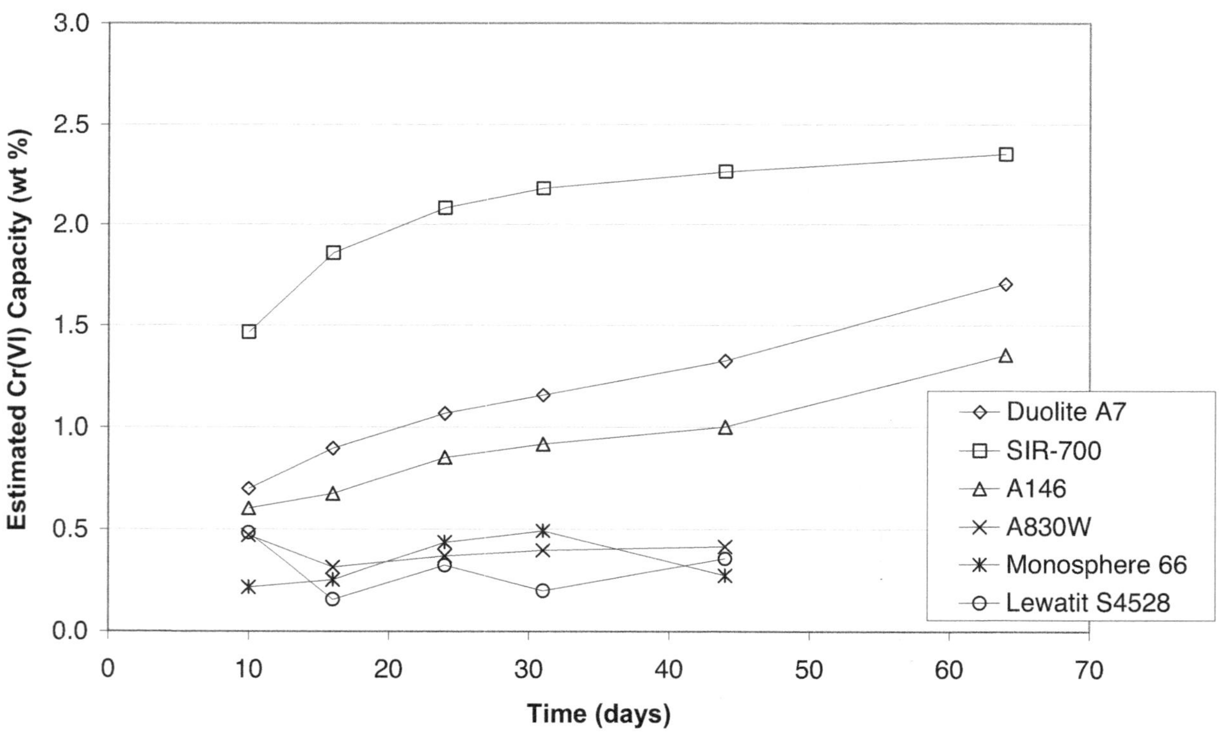

(b)

Figure 2.7 Estimated Cr(VI) capacity of the WBA resins for a resin dose of 40 mg/L at (a) pH 5.9, and (b) pH 6.4

Cr(VI) Removal Isotherms and Estimated Resin Capacities

Isotherm plots for the Duolite A7, SIR-700, and A146 resins on day 64 are shown in Figure 2.6 at pH values of 5.9 and 6.4. Although the timeframe for the three resins to achieve complete equilibrium was not specifically assessed in bench-scale testing, the aqueous phase Cr(VI) concentration on day 64 was judged to be close to the equilibrium concentration based on Figures 2.2 through 2.5.

As seen in Figure 2.6, the lowest Cr(VI) aqueous equilibrium concentrations and highest Cr(VI) resin removals were observed for SIR-700 resin, followed by Duolite A7, and then A146. Although fairly linear isotherms were observed for the three resins shown in Figure 2.6, residuals testing results described in Chapter 5 indicates that more than a traditional ion exchange mechanism was responsible for Cr(VI) removal by the WBA resins tested.

Contaminant removal capacities of the selected WBA resins were estimated based on the bench-scale isotherm testing. Calculated Cr(VI) capacities from one resin dose were plotted as a function of equilibrium time in Figure 2.7. SIR-700, Duolite A7, and A146 resins were shown to have an estimated capacity of between 1.5 and 2.5 weight percent (i.e., 15,000 to 25,000 μg of Cr per gram of resin). Figure 2.7 shows that a slightly reduced capacity might be expected at a water pH of 6.4 compared to 5.9 for the three resins.

Nitrosamine Leaching

Bench-scale nitrosamine testing was conducted to determine if any nitrosamines might be expected to leach from the top two performing resins (SIR-700 and Duolite A7). The functional groups on these WBA resins have not been revealed by the resin manufacturers at this time. These tests were run at the bench-scale after pilot testing due to nitrosamine analytical method unavailability at the time of the pilot testing.

Table 2.3 shows that two nitrosamines were detected in the leaching tests – NDMA and NPIP. The results shown in Table 2.3 indicate that initial flushing or preconditioning of the resins may be necessary to remove nitrosamines. Note, however, that another study of nitrosamine leaching from resins (Blute et al. 2006) indicated that the bench-scale leaching conditions may yield higher results compared with flow-through operations.

As of June 2007, California Department of Health Services (DHS) has Notification Levels for NDMA, NDEA, and NDPA of 10 ng/L. Notification Levels are not yet in published for NDBA, NPIP, NYPR, or NMEA. However, the California Office of Environmental Health Hazard Assessment (OEHHA) has reported low public health goal (PHG) levels for a number of nitrosamines without Notification Levels, establishing one-in-a-million cancer risk levels of 3 ng/L for NDBA, 3.5 ng/L for NPIP, 15 ng/L for NYPR, and 1.5 ng/L for NMEA. By comparison, one-in-a-million cancer risk levels for NDMA, NDEA, and NDPA are 3 ng/L, 1 ng/L, and 5 ng/L, respectively.

Note that the analytical method used in this testing, EPA Method 521, will be required for the upcoming Unregulated Contaminant Monitoring Regulation (UCMR2) but has not been approved by the EPA at any laboratories yet. As a result, these data are considered only tentative because of problems with the reliability of the method experienced by the contract laboratory. For example, the Duolite A7 sample had a 256% NDMA matrix spike recovery, compared to 77% for the GS-3 water and 97% for the SIR-700 sample. Consequently, these results present a qualitative assessment of what nitrosamines may leach from the resins rather than a quantitative analysis.

Table 2.3
Nitrosamine leaching from Duolite A7 and SIR-700 resins in bench-scale testing (ng/L)

Nitrosamine	GS-3 water (influent)	Duolite A7	SIR-700
NDMA	ND (< 2.0)	22	8.8
NDEA	ND (< 2.0)	ND (< 2.0)	ND (< 2.0)
NDPA	ND (< 2.0)	ND (< 2.0)	ND (< 2.0)
NDBA	ND (< 2.0)	ND (< 2.0)	ND (< 2.0)
NMEA	ND (< 2.0)	ND (< 2.0)	ND (< 2.0)
NPIP	ND (< 2.0)	42	ND (< 2.0)
NYPR	ND (< 2.0)	ND (< 2.0)	ND (< 2.0)

SUMMARY

The findings of the bench-scale evaluation to screen six promising WBA resins for Cr(VI) removal from Glendale groundwater are summarized below:

- Cr(VI) removal by three resins (SIR-700, Duolite A7, and A146) required more than 64 days to reach equilibrium, which was considered to be kinetically slow as compared to the removal of other anions using typical ion exchange resins.
- Cr(VI) removal by the other three resins (A830, Monosphere 66, and Lewatit S4528) was relatively fast, but their capacities for Cr(VI) removal were lowest.
- The near-equilibrium isotherm results showed that the SIR-700 and Duolite A7 resins had the two highest Cr(VI) removal capacities among the six WBA resins.
- High removal capacities of the SIR-700 and Duolite A7 resins, together with slow kinetics and residuals data (reported in Chapter 5), indicated that a mechanism other than ion exchange could be involved in Cr(VI) removal.
- Initial bench-scale testing of two pH levels (pH 5.9 and 6.4) indicated that slightly lower Cr(VI) capacities may be expected at the higher pH. Additional testing was needed to confirm the impact of pH on Cr(VI) capacity under flow-through conditions.
- Bench-scale testing indicated that the two top-performing resins may leach NDMA and NPIP; nitrosamines monitoring is recommended for full-scale operations and initial flushing or preconditioning of the resins may be necessary for nitrosamine removal.

CHAPTER 3
MINI-COLUMN EVALUATION

One question left unanswered in the bench-scale isotherm evaluation was the optimum pH for Cr(VI) removal by the selected WBA resins. pH can affect Cr(VI) removal by ion exchange resins in many ways. Theoretically, an acidic pH (4.5 to 5.0) may be preferred for Cr(VI) removal by WBA resins due to the occupation of more ion exchange sites per chromium atom by $HCrO_4^-$ at acidic pH compared to CrO_4^{2-} and less competition from OH^- for the exchange sites. WBA resins also require that a pH less than 6.5 be maintained so that the secondary or tertiary amine functional groups are protonated and thus act as positively charged exchange sites to attract anions (Clifford 1999). Reduction in pH, however, translates to a higher treatment cost from a drinking water treatment perspective (e.g., larger acid volumes and capital costs for acid storage). Mini-column testing was performed to identify the highest operating pH for WBA resins, such that the highest Cr(VI) capacity could be achieved with minimal acid feed costs.

EXPERIMENTAL METHODOLOGY

Mini-column testing was performed using five identical clear PVC columns 0.5-inch in diameter and 24-inch in length. Two resins were tested based on bench-scale testing: Duolite A7 and ResinTech SIR-700. Table 3.1 shows additional properties for these two resins beyond the basic information contained in Table 2.1.

Five different pH conditions (7.2, 6.8, 6.4, 6.0, and 5.6) were tested in mini-columns. The testing pH of 7.2 was selected based on historical laboratory water quality results of the GS-3 groundwater. The selection of pH 5.6 as the lowest testing pH was based on Phase II Pilot-Scale Study findings that below pH 5.6, Cr(III) might be released from the Duolite A7 resin (McGuire et al. 2006). For Glendale, a successful Cr(VI) treatment technology must not release Cr(III) above the treatment goal since Cr(III) can be re-oxidized to Cr(VI) by disinfectants in the distribution system (Brandhuber et al. 2004). Raw water from the GS-3 well was added to five 20-L overhead holding tanks, and then spiked with concentrated (32%) hydrochloric acid to reach the target pH values. The pH-adjusted water flowed through the mini-columns by gravity. The presence of 35 to 40 µg/L Cr(VI) in the GS-3 well enabled flow-through testing without Cr(VI) spiking. Figure 3.1 shows a photograph of the mini-column test system.

Operating conditions and column specifications were developed based on previous testing experience and resin manufacturer recommendations. Table 3.2 lists the specifications for the mini-column testing. Based on these operating conditions, the water consumption rate for each column was approximately 18.5 L per day; hence, the water in the overhead tanks was replaced daily during the 10-day testing period. Although the testing period was not long enough for the development of complete breakthrough curves, the appearance of Cr(VI) concentrations in the effluent above the treatment goal was used to assess appropriate pH values to test in the subsequent pilot-scale column tests.

Influent and five effluent samples were collected daily during the mini-column evaluation. Total Cr and Cr(VI) analyses were performed at Utah State University for both the mini-column and pilot-scale column evaluations. Total Cr concentrations were measured using an Agilent 7500C inductively coupled plasma mass spectrometer (ICP-MS) with an octopole reaction system, in accordance with USEPA Method 200.8 (USEPA 1994). Cr(VI) concentrations were

Table 3.1
Resin properties

Property	Duolite A7	ResinTech SIR-700
Resin type	Weakly basic anion	Weakly basic anion
Resin physical form	Cream colored granules	Yellow granules
Particle size	0.3–1.2 mm (16–50 mesh US Std screen)	0.3–1.7 mm (12–50 mesh US Std screen)
Suggested hydraulic loading rate	2–10 gpm/ft^2	2–8 gpm/ft^2
Uniformity coefficient	Less than 2.0	Less than 2.0

NOTES:
1. The cartridge filter, pressure gauge, and data recorder in the shaded area is not part of the mini-column test system.
2. Four of the five overhead holding tanks are shown in the picture.

Figure 3.1 Photograph of the mini-column test system

Table 3.2
Mini-column specifications and operating conditions

Parameter	Specification
Column diameter	0.5 inches
Resin bed depth	8 inches
Bed volume (BV)	0.0009 cubic feet (0.007 gallon)
Operating pH	7.2, 6.8, 6.4, 6.0, 5.6
Operational mode	Downflow
Hydraulic loading rate	2.5 gpm/ft^2
Empty bed contact time (EBCT)	2 min.
Raw water flow rate	0.0034 gpm (13 mL/min)
Estimated bed volumes treated	720 BV/day
Estimated water consumption per column	4.9 gallons/day (18.5 L/day)
Run time of each column	10 days
Backwash capability	None

measured using a Dionex DX-320 ion chromatograph (IC) with an AD25 post-column ultraviolet-visible detector according to USEPA Method 1636 (USEPA 1996). Field analysis of pH was conducted using an an Accumet AB+ benchtop pH meter (Fisher Scientific Company, Pittsburgh, Pa.).

RESULTS AND DISCUSSION

Observed pH Change

During the first few days of the mini-column evaluation, a dramatic pH increase (from 0.3 to 0.8 pH units) inside the overhead holding tanks was observed over the course of 24 hours. A raw water sample collected from the GS-3 well was exposed to the atmosphere (similar to the overhead tank set-up) and the pH increased from approximately 7.0 to 8.1 over a 72-hr period (Figure 3.2), most likely due to the release of supersaturated carbon dioxide from the groundwater. Since constant pH levels were critical in the WBA resin testing, the mini-column evaluation was temporarily suspended while an aeration system was put in place to stabilize the influent pH by stripping carbon dioxide. Using the aeration system, the GS-3 feed water pH could be stabilized at pH 8.3 after 15 minutes. This stabilized feed water was then added to each holding tank and adjusted to the desired pH values. After the modification, the pH values in the holding tanks were more stable, as indicated by an increase of approximately 0.2 pH units overnight. It was determined that such increase was acceptable and the mini-column evaluation was resumed.

Of note from these results is that the groundwater pH subsequently determined using an in-line pH probe under pressure was around 6.8 – i.e., lower than the historical laboratory results indicating pH 7.2. Treatment cost savings would thus be realized if the WBA system were operated without breaking head since less acid would be needed to depress the pH.

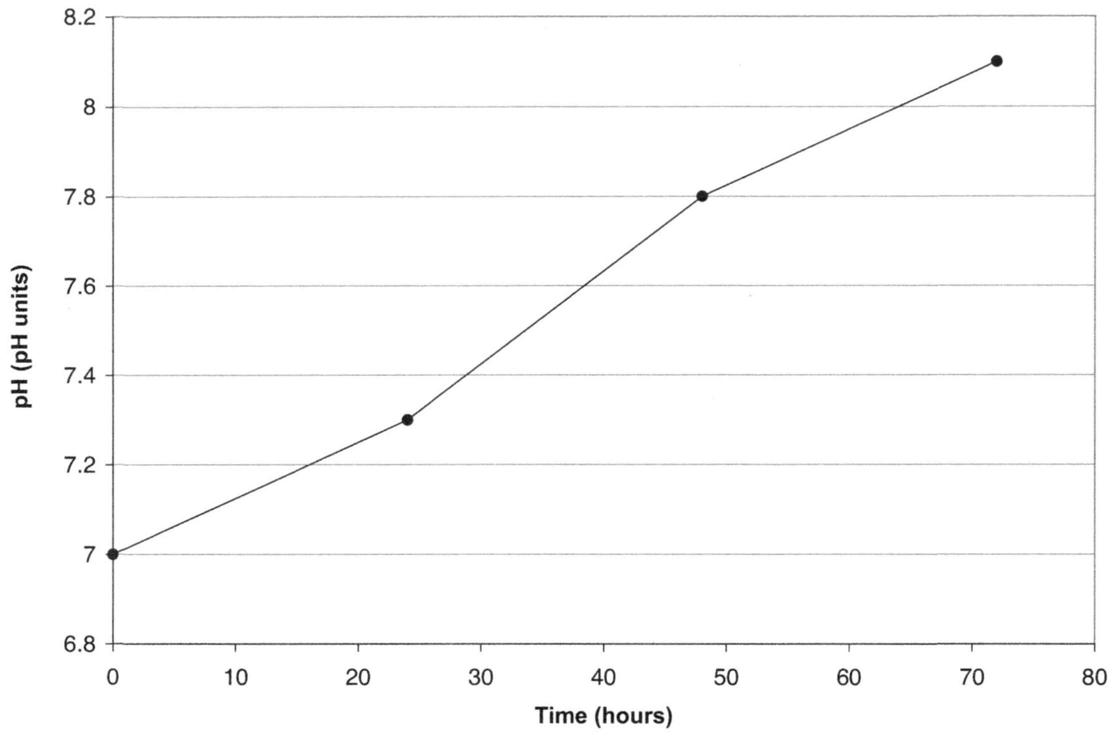

Figure 3.2 pH increase in the GS-3 well water over 72 hours

Impact of pH on Chromium Removal

The mini-column evaluation was restarted once the feed water pH was stabilized. Figures 3.3 and 3.4 show the total Cr and Cr(VI) effluent concentrations after treatment by the Duolite A7 resin at each tested pH level. A pH of between 5.6 and 6.4 was determined to likely provide sufficient pH reduction for effective Cr removal by the Duolite A7 resin, as demonstrated by less than 1 μg/L of total Cr and Cr(VI) concentrations in the column effluents after treating 5,000 bed volumes (BV) of water.

Figures 3.5 and 3.6 show total Cr and Cr(VI) removals by the SIR-700 resin at each pH level tested in mini-columns. Different behavior was observed for the SIR-700 resin compared to the Duolite A7 resin. The onset of breakthrough in Figure 3.5 demonstrated that SIR-700 resin operated with less Cr(VI) leakage at an operating pH of 6.0 or lower.

In mini-column testing, speciation between Cr(VI) and Cr(III) in column effluents was difficult to quantify due to limited data and effluent concentrations less than or near the method detection limit (i.e., 0.4 μg/L for Cr(VI) and 0.1 μg/L for total Cr). However, the similar patterns between Figure 3.3 and Figure 3.4 and between Figure 3.5 and Figure 3.6 demonstrated that Cr(VI) was the major chromium species in the mini-column effluents.

Since comparable total Cr and Cr(VI) removals were observed at pH 6.0 and 5.6 for both resins, the higher pH of 6.0 was chosen for subsequent pilot-scale column testing in an effort to lessen possible full-scale acid addition costs. Because pH 6.8 was found to be the ambient pH and Duolite A7 had not exceeded the treatment goal of 5 μg/L in the mini-column testing at this pH, pilot-scale testing also included pH 6.8.

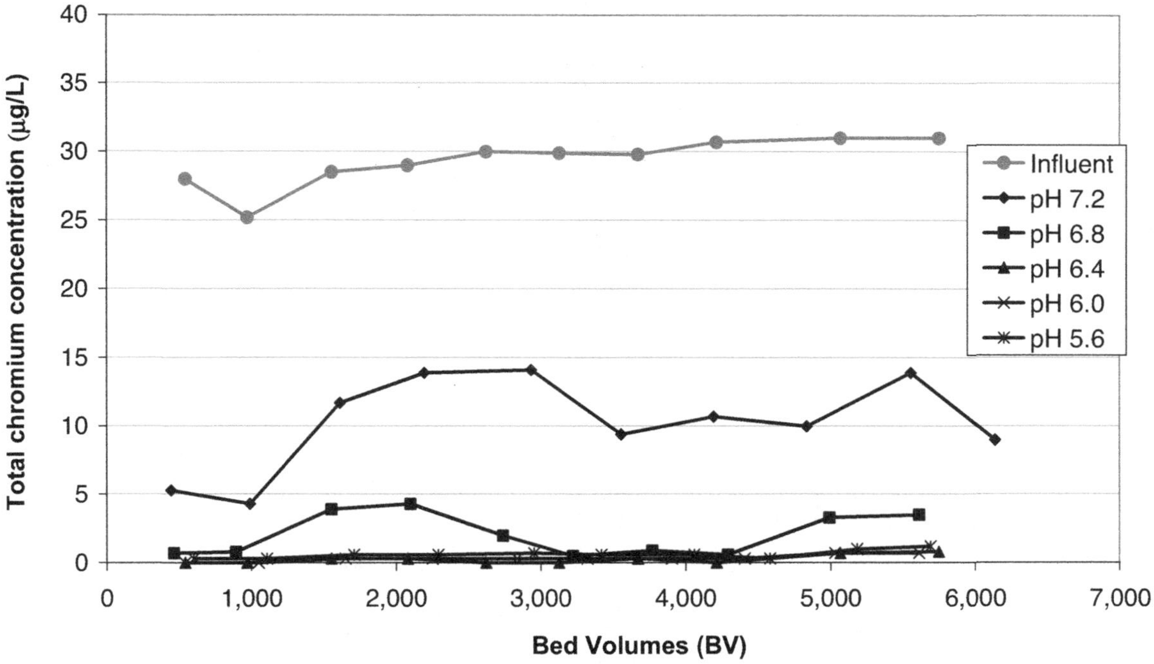

Figure 3.3 Mini-column evaluation of Duolite A7 resin at different pH values – Total Cr removal

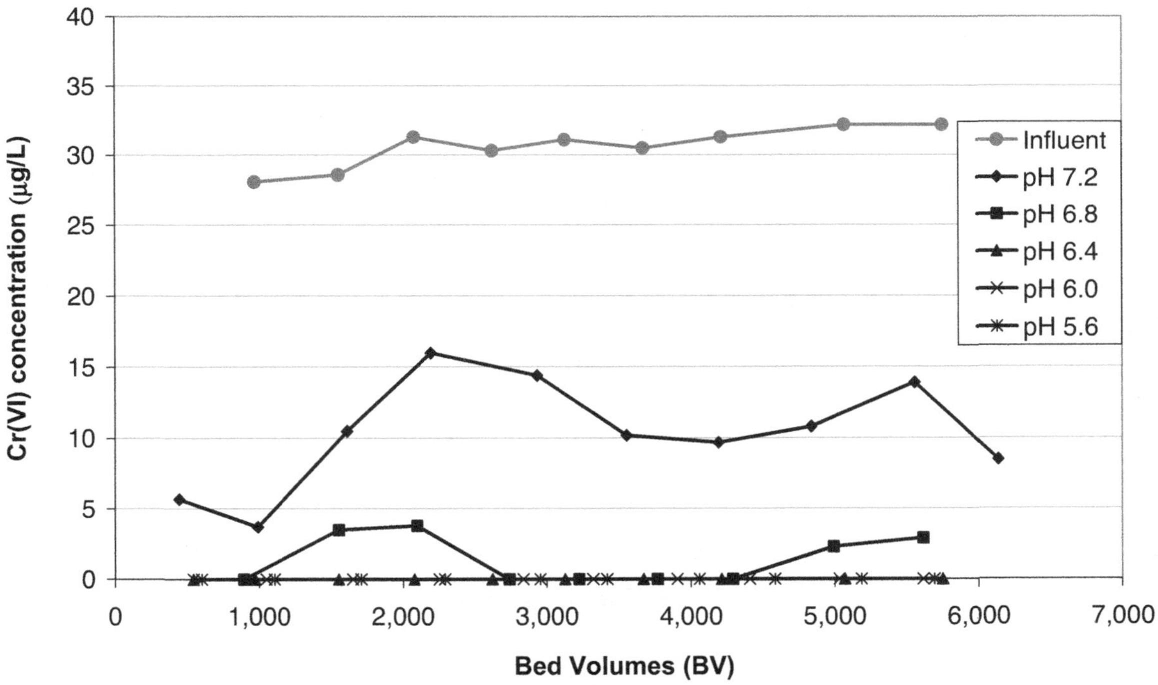

Figure 3.4 Mini-column evaluation of Duolite A7 resin at different pH values – Cr(VI) removal

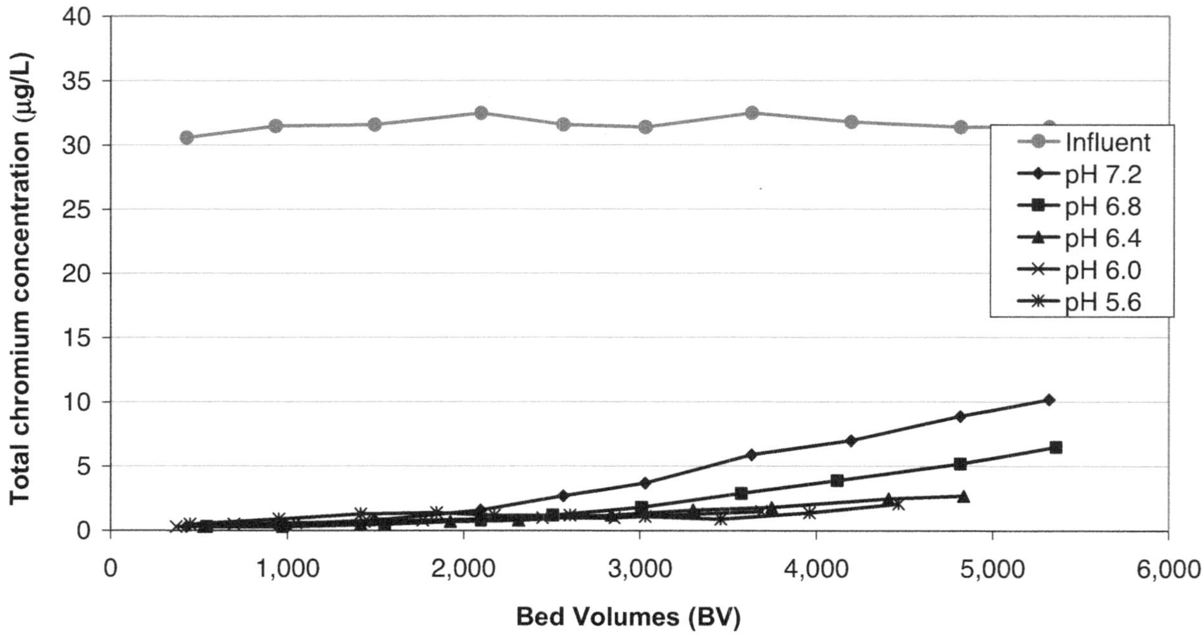

Figure 3.5 Mini-column evaluation of SIR-700 resin at different pH values – Total Cr removal

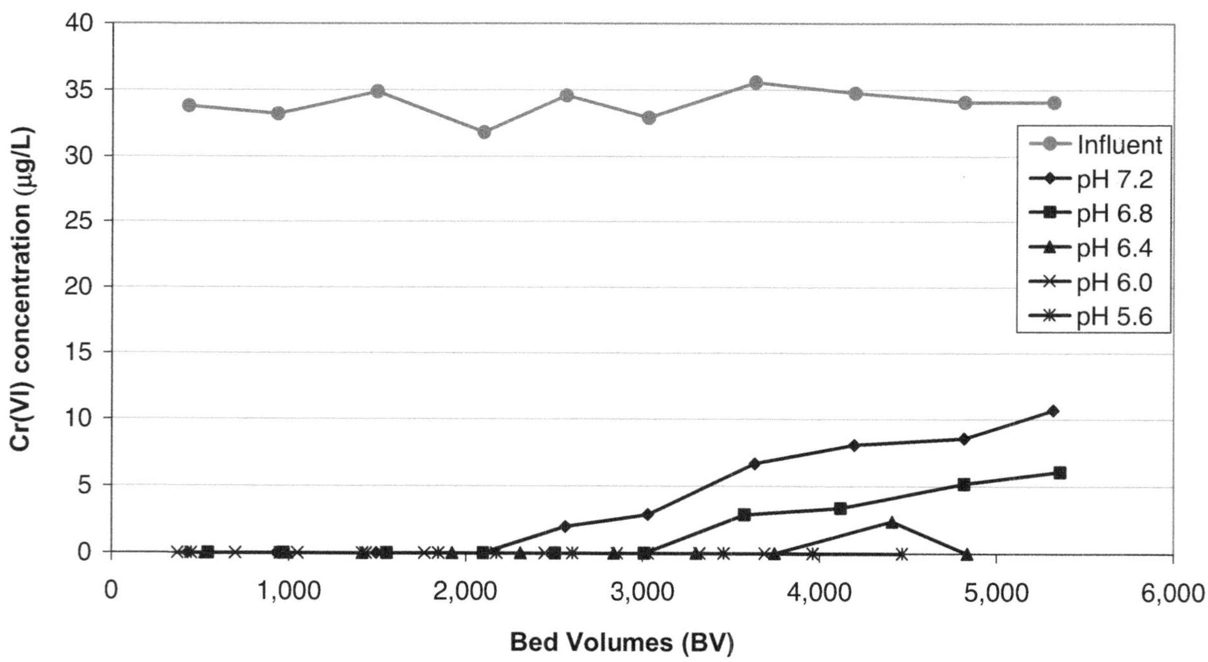

Figure 3.6 Mini-column evaluation of SIR-700 resin at different pH values – Cr(VI) removal

SUMMARY

The two top-performing WBA resins in the bench-scale isotherm evaluation, Duolite A7 and SIR-700, were further tested in mini-columns at five different pH conditions. The findings from the mini-column evaluation are summarized below:

- The water from the Glendale GS-3 well was supersaturated with CO_2 and had a natural pH around 6.8, which was less than the historical laboratory-measured pH data of 7.2. The lower influent pH would translate into a lower acid addition cost if the WBA system was operated without breaking head.
- Speciation between Cr(VI) and Cr(III) in column effluents for both resins was difficult to quantify in mini-column testing because significant breakthrough had not occurred. However, when detectable, Cr(VI) was the major chromium species measured in the column effluents.
- Mini-column testing results at pH values ranging from 5.6 to 7.2 showed different breakthrough characteristics for SIR-700 and Duolite A7 resins. Nonetheless, both resins favored a lower operating pH for total Cr and Cr(VI) removal.
- A pH of 6.0 was selected for pilot-scale testing, along with ambient pH 6.8, since Duolite A7 showed some promise of Cr(VI) capacity at the higher pH.

CHAPTER 4
PILOT-SCALE EVALUATION

Based on bench-scale and mini-column results, Duolite A7 and SIR-700 resins were selected for testing in the larger pilot-scale columns at constant pH values of 6.0 and 6.8. The purpose of pilot-scale column testing was to develop chromium breakthrough curves that would provide an estimate of resin capacity for full-scale WBA treatment. The impact of the resins on effluent water quality was also examined.

EXPERIMENTAL METHODOLOGY

The system for pilot-scale testing consisted of the following major components: four identical 2.5 inch diameter PVC columns, a storage tank and chemical feed pump for hydrochloric acid (HCl), an in-line pH sensor, flow meters, and a data recording system for pH and flow rates. The use of multiple columns enabled the evaluation of chromium treatment by two WBA resins at pH 6.0 (selected based on mini-column evaluation results) and at the groundwater ambient pH (approximately 6.8, without any acid injection). The schematic of the pilot system is shown in Figure 4.1 and a system picture is shown in Figure 4.2.

Originally, the four-column design was intended to simultaneously test two WBA resins at two pH conditions. However, a constant pH could not be maintained with the original design due to the lack of adequate mixing of the acid and influent water by the static mixer. The pilot system was modified so that one column was used as a mixing column between the acid injection point and the on-line pH probe, providing an extra 2 minutes of mixing time prior to pH measurement. The pH control of the pilot system was greatly improved after the modification, as indicated by pH variations of less than a 0.2 pH units typically observed over 24-hr periods.

On-site analyses of influent and effluent water quality were conducted during the pilot-scale column evaluation to assess the impact of resin operation on effluent water quality. pH was measured throughout the testing using two types of probes: an on-line pH probe (Signet 2714 pH electrode from George Fischer Signet, Inc., Calif.) and a benchtop Accumet AB+ pH meter (Fisher Scientific Company, Pittsburgh, Pa). The on-line pH probe measured the pressurized influent pH after acid addition and mixing, while the field pH meter was used to measure pH in grab samples of influent (after acid addition) and the pH 6.0 test column effluents.

The pilot-scale column specifications and operating conditions are summarized in Table 4.1. Key operating conditions, such as EBCT and hydraulic loading rate, were selected based on conversations with the resin manufacturers and the desire to test the resins under conditions within the range that would be recommended for full-scale treatment (i.e., 2 to 10 gpm/ft^2).

Since only three columns were available for resin testing at any time, the two resins were sequentially tested under ambient pH conditions. Given the relatively stable water quality in the GS-3 well, the performance of two resins tested sequentially was not expected to differ significantly from parallel testing conditions.

Influent and effluent samples were collected once or twice a week for total Cr and Cr(VI) analyses at USU. One influent sample was collected upstream of flow splitting into the three columns and also acid addition. Field measurements of pH were obtained at the same frequency as the chromium analyses. Other water quality parameters in influent and effluent samples,

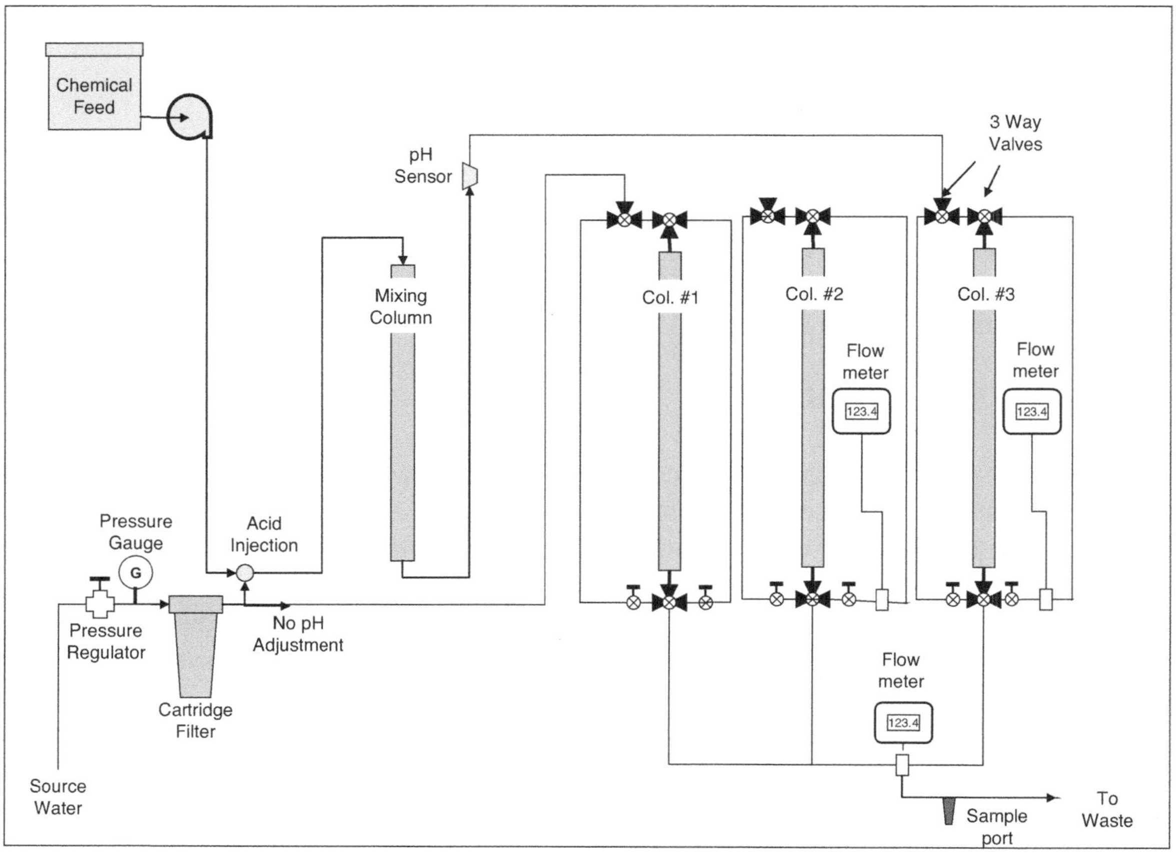

Figure 4.1 Schematic diagram of the pilot-scale test system

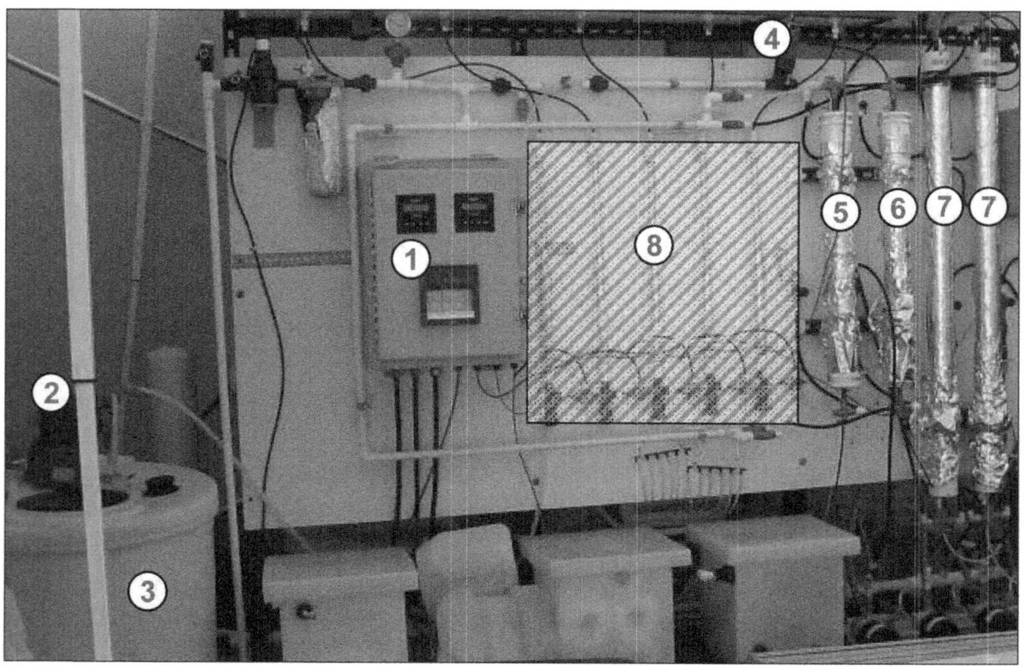

Key: 1. Control box and data recorder; 2. Acid feed pump; 3. Acid feed solution; 4. pH probe; 5. Mixing column; 6. Ambient pH test column; 7. pH 6.0 test columns; 8. Mini-columns (not part of the pilot testing)

Figure 4.2 Photograph of the pilot-scale test system

Table 4.1
Pilot-scale column specifications and operating conditions

Parameter	Specification
Column diameter	2.5 inches
Resin bed depth	13 inches
Bed volume (BV)	0.036 cubic feet (0.27 gallons)
Operating pH	6.0 and ambient pH (approximately 6.8)
Operational mode	Downflow
Hydraulic loading rate	4 gpm/ft^2
Service flow rate	3.9 gpm/ft^3
EBCT	2 min
Raw water flow rate	0.14 gpm (516 mL/min)
Estimated bed volumes treated	720 BV/day
Estimated water consumption per column	196 gallons/day (743 L/day)
Run time	Until breakthrough is observed
Backwash frequency	Backwash when severe head loss is observed

Note: No backwashing was necessary during the pilot testing.

including sulfate, nitrate, phosphate, conductivity, turbidity, alkalinity, and hardness, were monitored periodically using Hach field instruments.

RESULTS AND DISCUSSION

Chromium Removal by WBA Resins

Total Cr and Cr(VI) breakthrough curves for Duolite A7 and SIR-700 resins at pH 6.0 and ambient pH are depicted in Figures 4.3 and 4.4. At pH 6.0, effluent from the Duolite A7 column reached a total Cr concentration of 5 μg/L at approximately 45,000 BV. More than 113,000 BV (113,000 L) of water was treated in the pilot testing, yielding an effluent concentration approaching 15 μg/L at the end. Total Cr and Cr(VI) breakthrough curves of the Duolite A7 resin operated at pH 6.0 were shown to rise slowly, indicating a long operational life might be achieved if the resin was used in a lead-lag configuration. In contrast to the pH 6.0 case, Duolite A7 resin operated at ambient pH yielded effluent total Cr concentrations above 5 μg/L after 2,300 BV, which gradually increased to 25 μg/L at 80,000 BV. Based on these results, a lower pH was shown to be necessary to achieve Cr(VI) removal to levels below the target treatment goal for an extended period of time. The consistent chromium removal performance over the testing period and resin capacity indicated that the Duolite A7 resin had the potential of being used successfully in a single-pass mode for Cr(VI) removal from Glendale groundwater.

For the SIR-700 resin operating at ambient pH, the column effluent exceeded the 5 μg/L total Cr target at approximately 1,800 BV and rapidly increased to about 25 μg/L at 7,000 BV. Operation of the SIR-700 resin at ambient pH was terminated due to the near complete exhaustion

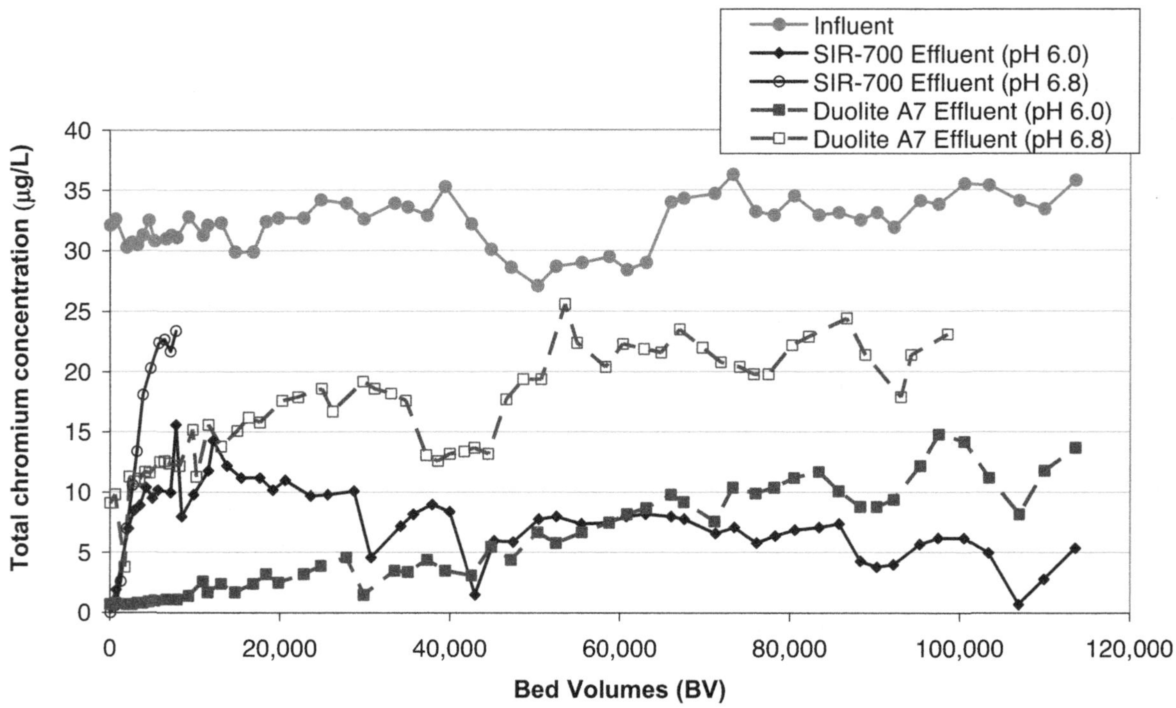

Figure 4.3 Pilot-scale column breakthrough curves of total Cr at pH 6.0 and ambient pH for Duolite A7 and SIR-700 resins

of the media. At pH 6.0, the Cr(VI) and total Cr breakthrough curves were quite similar to the curves at ambient pH for the first 2,000 BV. Effluent concentrations above 5 µg/L total Cr were also observed at approximately 1,800 BV. However, after reaching 15 µg/L at 12,000 BV, the total Cr effluent concentrations slowing decreased to less than 5 µg/L. The decrease from 15 µg/L to approximately 5 µg/L was beyond the normal variability of the analytical results, because the same pattern can be observed in Figures 4.3 and 4.4 [i.e., Cr(VI) and total Cr concentrations were measured with different analytical methods]. The early breakthrough of chromium from the SIR-700 resin might limit its application in Glendale groundwater. Nonetheless, the decrease in the breakthrough curve as a function of bed volumes of water treated indicated that chromium removal by the SIR-700 resin was not reflective of classic ion exchange mechanism. A better understanding of its removal mechanism, and potential resin pre-conditioning, might make the SIR-700 resin more suitable for Cr(VI) treatment of drinking water.

As observed in Figures 4.3 and 4.4, the dominant chromium species in the influent and effluent samples was Cr(VI). Previous Phase II Pilot-Scale Study results, however, suggested that Cr(III) leakage may be observed at lower pH values than tested in Phase III (generally less than 5.5; McGuire et al. 2006). The appearance of Cr(III) in the Phase II testing was surprising since Cr(VI) is the major chromium species in the source water. Typical anion exchange processes do not involve redox changes with anions removed from the aqueous phase. However, the Phase II results suggested that Cr(VI) was reduced to Cr(III) through a redox reaction with the resin. As described in Chapter 5, the Phase III Bridge Project included a component to probe the mechanism of Cr retention by the WBA resins.

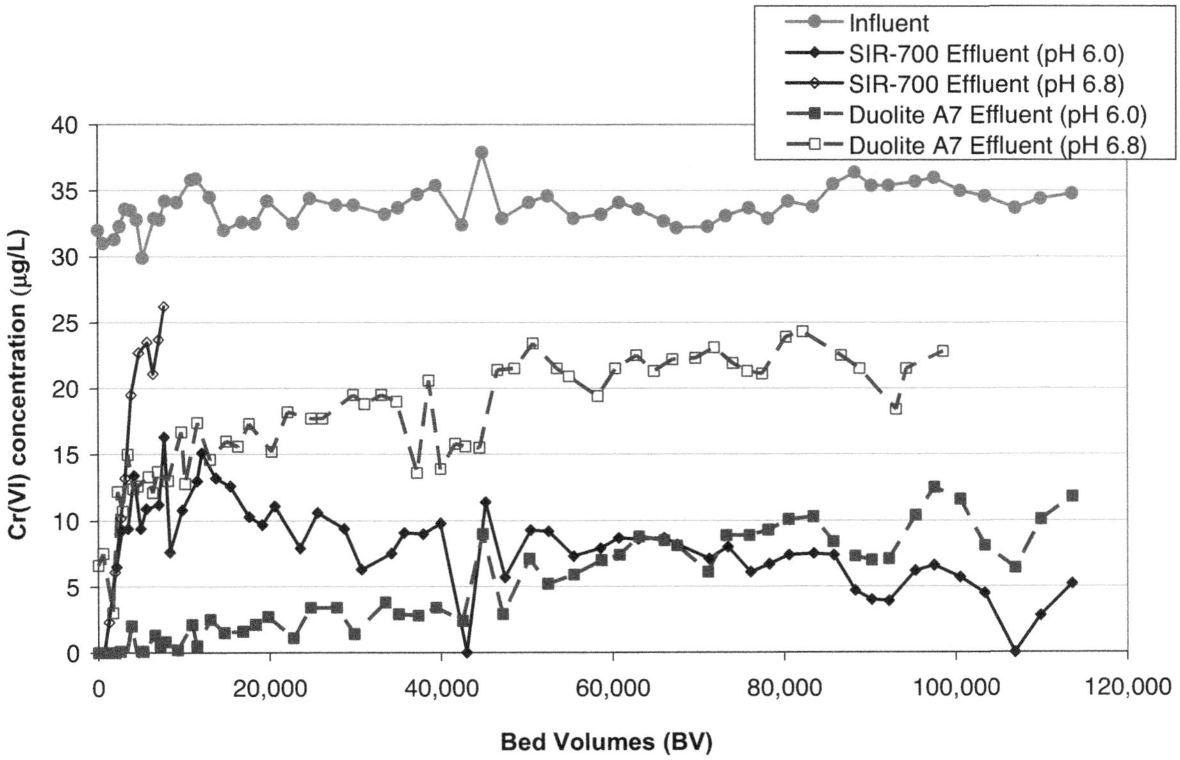

Figure 4.4 Pilot-scale column breakthrough curves of Cr(VI) at pH 6.0 and ambient pH for Duolite A7 and SIR-700 resins

Effluent Water Quality

Figure 4.5 shows the influent water pH profiles as measured by the on-line probe and the field meter, as well as the pH 6.0 column effluents measured by the field meter. As illustrated in Figure 4.5, the influent pH measured by field pH meter was always higher than the influent pH measured using the on-line probe due to the offgassing problem discovered during the mini-column evaluation (refer to Chapter 3 for details). A comparison of the pH values between influent and effluent samples using the field pH meter showed little change in pH due to the flow of water through the resin.

In addition to pH, other parameters in influent and effluent samples were monitored during the pilot testing. Table 4.2 lists the average concentrations of the water quality parameters in influent (without acid addition) and effluent samples. With the exception of alkalinity, the impact of the WBA resin operation on water quality was minimal. The reduction in alkalinity in Duolite A7 and SIR-700 effluents at pH 6.0 was likely caused by acid addition.

SUMMARY

Two WBA resins, SIR-700 and Duolite A7, were tested using pilot-scale columns to assess Cr(VI) capacity under typical operational conditions. Two different pH values were tested (pH 6.0 and 6.8). Findings from the pilot-scale column evaluation are summarized below:

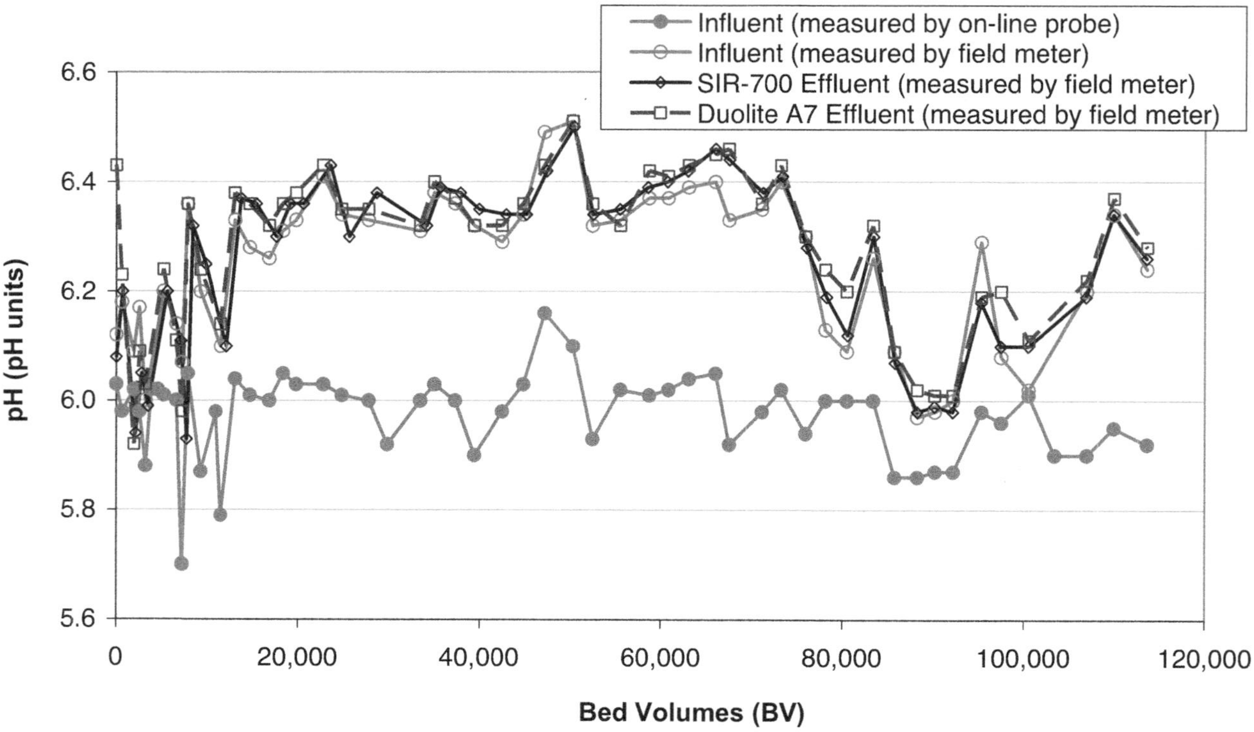

Figure 4.5 pH profiles of the influent and the pH 6.0 test column effluents

Table 4.2
Water quality parameters of influent and column effluents during the pilot-scale column evaluation

	Influent (without acid addition)	Duolite A7 effluent (pH 6.0 test column)	Duolite A7 effluent (ambient pH test column)	SIR-700 effluent (pH 6.0 test column)	SIR-700 effluent (ambient pH test column)
NO_3_N (mg/L)	6.5	6.7	5.1	6.4	6.3
SO_4^{2-} (mg/L)	112	111	125	114	117
PO_4^{3-} (mg/L)	0.27	0.25	0.28	0.27	0.29
SiO_2 (mg/L)	33	33	33	32	33
Conductivity (µS/cm)	848	858	786	860	796
Alkalinity (mg/L as $CaCO_3$)	194	82	194	80	186
Hardness (mg/L as $CaCO_3$)	350	348	352	356	373
Turbidity (NTU)	0.1	0.1	0.1	0.1	0.1

- Both SIR-700 and Duolite A7 resins achieved better chromium removal at pH 6.0 compared to ambient pH.
- At pH 6.0, Duolite A7 resin was shown to remove total Cr and Cr(VI) to less than 5 μg/L for approximately 45,000 BV and to 50% breakthrough (approximately 15 μg/L) for more than 113,000 BV.
- The consistent removal of Cr(VI) and total Cr to below 5 μg/L by the Duolite A7 resin during pilot testing indicated the resin had the potential of being used in a single-pass mode for Cr(VI) removal in Glendale groundwater.
- SIR-700 resin exhibited 5 μg/L breakthrough at 2,200 BV for pH 6.0. However, Cr(VI) removal appeared to improve during the testing period, which is contrary to the typical anion exchange mechanism and may suggest another removal mechanism.
- Cr(VI) was the primary chromium species in both influent and effluent samples.
- The impact of WBA resin operation on effluent water quality was shown to be minimal for the parameters measured, with the exception of NDMA and NPIP leaching.

CHAPTER 5
RESIDUALS ANALYSES

Residuals management is one of the most important drivers for water treatment process selection because it can significantly impact full-scale operational costs and staffing requirements. As such, assessment of residuals minimization and disposal options for the WBA resins was a key aspect of the Phase III Bridge Study.

The mechanism of Cr(VI) removal by the WBA resins was also explored through the analysis of resin residuals. As mentioned previously, initial pilot testing results from the Phase II Pilot-Scale Study indicated that Cr(VI) reduction might occur in the resin columns. On occasions in which influent pH levels dropped below 5.5, Cr(III) (at concentrations significantly higher than the influent concentration) was released from the columns. Around the same time, another group's pilot investigation of Duolite A7 resin surmised that some part of the resin might become oxidized and reduce Cr(VI) to Cr(III) in the process (Höll et al. 2002). The pilot-scale column evaluation presented in Chapter 4 suggested that ion exchange alone did not appear to explain the Cr(VI) removal performance by the WBA resins.

EXPERIMENTAL METHODOLOGY

Once the pilot-scale column evaluations were complete, spent Duolite A7 and SIR-700 resins were removed from columns and analyzed using a series of methods. Note that the resins were operated to different endpoints based on as listed below depending on effluent chromium concentrations:

- Duolite A7 pH 6.0 – 113,594 BV water treated
- SIR-700 pH 6.0 – 113,566 BV water treated
- Duolite A7 pH 6.8 – 98,538 BV water treated
- SIR-700 pH 6.8 – 7,731 BV water treated

The resin residuals from the columns operating at pH 6.0 were tested for hazardous waste characteristics using the federal toxicity characteristic leaching procedure (TCLP; USEPA 1998) and the California Waste Extraction Test (WET; Office of Administrative Law, 1991). Total uranium concentrations on the spent resins were measured by kinetic phosphorescence analysis (KPA) in accordance with ASTM 5174-91 (Severn Trent Laboratories, Earth City, Mo.).

In addition to the hazardous waste characterization analyses, other geochemical methods were also used to analyze the resin residuals. With the exception of one test column, spent resin samples were evenly divided into three groups according to their location inside the testing column: top one-third of the column, middle one-third, and bottom one-third. For SIR-700 resin at pH 6.8 that was finished before the other three columns, the resin was divided into two sections rather than three. Subsequently, the decision was made to split the remaining resin columns into thirds.

Cr speciation on resin subsamples was determined using x-ray absorption near-edge structure (XANES) spectroscopy. Previous studies have demonstrated the feasibility of determining Cr speciation on solid samples using the XANES spectroscopy (O'Day et al. 2000, Bond and Fendorf 2003, Berry and O'Neill 2004, Wilkin et al. 2005). X-ray absorption spectra of the spent

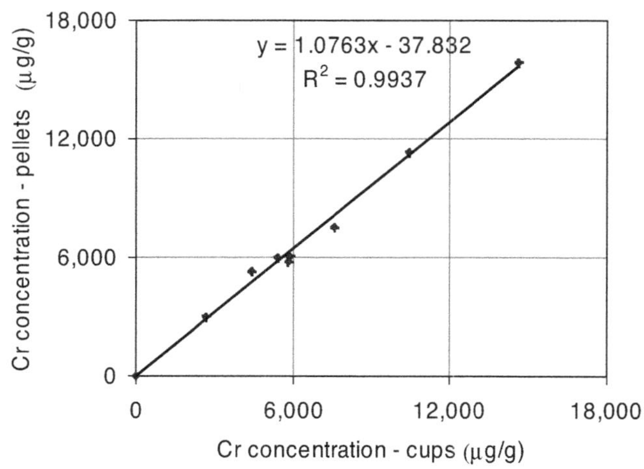

Figure 5.1 Total Cr concentrations measured by XRF using pellets or cups

resins were collected at beamline 13-BM-D at the Advanced Photon Source, Argonne National Laboratory (Argonne, Ill.). Resin samples were loaded onto the sample holder and scanned by the incident X-ray beam near the Cr K-edge region (5989 eV for metallic Cr). The absorption spectra were collected in fluorescence mode (rather than transmission mode) due to the thickness of the sample. In order to monitor any oxidation state change caused by the high energy beamline, the resin samples were first scanned for varying durations (i.e., from seconds to minutes). The lack of changes in the spectra provided evidence that the chromium oxidation states were not changed by the incident x-ray beam. Reference compounds, including Cr(0) metal foil, Cr(III) compounds (chromium acetate, chromium oxide, chromium nitrate, and chromium ferrous oxide), and a Cr(VI) compound (ammonium dichromate), were also analyzed using the same beamline. Each sample or reference compound was scanned 2–3 times to reduce spectral noise. The XANES spectral analysis was performed using the computer program ATHENA (Ravel and Newville 2005), with functions including spectra averaging and merging, background removal, and data normalization.

Total concentrations of chromium and other elements retained on the spent resins were determined with an XEPOS x-ray fluorescence (XRF) spectrometer (Spectro Analytical Instruments, Inc., Marlborough, Mass.) at Wellesley College (Wellesley, Mass.). Sample preparation for this technique typically involves grinding the material into a fine powder and pressing the material into a pellet under high pressure. However, the nature of the resin material did not allow several of the resin samples to be pressed into a pellet (or pellets broke in the XRF sampling tray). Instead, resin samples were finely ground and packed into a plastic cup with a plastic film for analysis of that surface. Figure 5.1 shows a comparison of some of the resins that were able to be made into pellets and the same resins using the plastic cup. The figure indicates that the cup method yields similar results as the pellet method for chromium.

Resin subsamples were also analyzed by x-ray diffraction (XRD) to determine if a dominant crystalline phase was present. XRD was also conducted at Wellesley College using a Rigaku 300 diffractometer (Rigaku, The Woodlands, Texas) with a rotating copper anode, and the data were analyzed using the computer program Jade.

Finally, resin samples were analyzed by scanning electron microscopy (SEM) to determine if chromium was evenly distributed in the spent resin or present as a precipitate on the

Table 5.1
TCLP metals analysis results on spent resins

	Rohm & Haas Duolite A7 (pH 6.0)	ResinTech SIR-700 (pH 6.0)	Regulatory limit
Arsenic	ND	ND	5,000
Barium	ND	ND	100,000
Cadmium	ND	ND	1,000
Chromium	260	45	5,000
Lead	24	15	5,000
Mercury	ND	ND	200
Selenium	ND	ND	1,000
Silver	ND	ND	5,000

NOTE: Concentrations in the table are expressed in μg/L. ND = non-detect.

surface. SEM imaging was conducted at the Massachusetts Institute of Technology using a LEOVP438 with an iXRF energy dispersive analytical system. All images are backscatter electron (BSE) mode collected with a chamber pressure of 10 Pascals, a probe current of 250 pA, and an excitation voltage of 20kV at 19 mm working distance.

RESULTS AND DISCUSSION

Hazardous Waste Characterization

Results of the Federal TCLP extraction test (metals analysis) and California WET test [soluble threshold limit concentration (STLC) metals analysis] on spent SIR-700 and Duolite A7 resins are shown in Tables 5.1 and 5.2, respectively. The TCLP and WET analyses were not performed on the virgin resins since the metal concentrations in the virgin resins were assumed to be far below the regulatory limits. This assumption was later supported by the XRF results.

In the TCLP test, only chromium and lead were leached out from spent Duolite A7 and SIR-700 resins, and the concentrations were far less than the regulatory limit. Therefore, both resins passed the TCLP test and would not be considered as hazardous waste by federal standards.

More metal species were leached out during the California WET test than the TCLP test. Chromium concentrations in the WET test leachate were 11,200 and 10,400 μg/L for Duolite A7 and SIR-700 resin residuals, respectively. Both resins would be considered as hazardous waste for disposal in California since the WET testing chromium limit is 5,000 μg/L. It is worth noting that the copper concentration in the Duolite A7 resin leachate was 22,400 μg/L, which is approaching the regulatory limit of 25,000 μg/L. The relatively high concentration of copper determined from the WET test on both resins was an unexpected finding. Relatively high concentrations of vanadium were also detected in the resin leachates.

Figure 5.2 shows the total uranium levels accumulated on the spent WBA resins. The uranium concentration on the spent Duolite A7 resin was 536 μg/g, exceeding the 500 μg/g limit

Table 5.2
California WET metals analysis (STLC) results

	Rohm & Haas Duolite A7 (pH 6.0)	ResinTech SIR-700 (pH 6.0)	Regulatory limit
Antimony	ND	ND	15,000
Arsenic	ND	ND	5,000
Barium	29	ND	100,000
Beryllium	3	ND	750
Cadmium	ND	ND	1,000
Chromium	11,200	10,400	5,000
Cobalt	ND	ND	80,000
Copper	22,400	13,300	25,000
Lead	ND	ND	5,000
Mercury	ND	ND	200
Molybdenum	116	116	350,000
Nickel	52	ND	20,000
Selenium	ND	ND	1,000
Silver	ND	ND	5,000
Thallium	ND	ND	7,000
Vanadium	3,270	3,690	24,000
Zinc	443	81	250,000

NOTE: Concentrations in the table are expressed in μg/L. ND = non-detect.

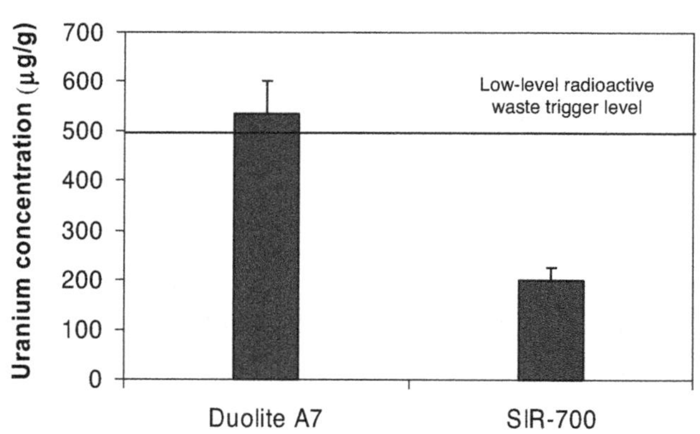

Figure 5.2 Total uranium concentrations on the spent WBA resins (resins run at pH 6.0)

(the trigger for low-level mixed radioactive waste designation) at the end of pilot testing (approximately 113,000 BV of water treated). In order to avoid the generation of a low-level mixed radioactive waste for source waters containing 1.4 pCi/L uranium, the operating life of the Duolite A7 resin would have to be limited to approximately 100,000 BV. The uranium concentration on spent SIR-700 resin (200 μg/g) was much lower than the regulatory limit and was, therefore, not considered to be a limiting factor for the SIR-700 resin's operation.

Solid-Phase Analyses

Cr XANES Spectra

Several observations from the related WBA resin studies suggested that the high Cr(VI) capacities of the WBA resins operating at pH 6.0 could not be explained by the ion exchange mechanism alone, including (1) the leaching of Cr(III) in the Duolite A7 effluent in the Phase II Pilot-Scale Study, (2) the shape of SIR-700 Cr breakthrough curve at pH 6.0, and (3) the change of SIR-700 resin color from the bright yellow color of virgin resin to the dark green color of spent resin (note that dark green is a characteristic color of Cr(III) compounds). An oxidation-reduction (redox) mechanism was, therefore, hypothesized to play a role in Cr(VI) removal by the WBA resins under slightly acidic condition.

The chromium XANES spectra for various reference materials are shown in Figure 5.3. The common feature shared by compounds with the same chromium oxidation state is the edge energy. The absorption edge of the chromium-bearing reference compounds is 5989 eV for Cr(0) (metal), 6003–6004 eV for Cr(III), and 6007 eV for Cr(VI). Due to d-p orbital hybridization, the Cr(VI) spectrum also has a sharp pre-edge absorption peak at approximately 5993 eV, which is absent in Cr(III) spectra.

The middle third of the spent Duolite A7 and SIR-700 resins are compared with selected known Cr reference materials in Figure 5.4. The edge overlap between the spent resin samples and Cr(III) reference compounds and the lack of any pre-edge peak indicated that Cr(III) was the dominant species (i.e., comprising more than 95% of the chromium) retained on both spent resins. This finding provided direct evidence that Cr(VI) was reduced to Cr(III) by both WBA resins.

XRF Results

The accumulation of total chromium and other compounds on the spent resins was determined using XRF. Concentrations for various resin depths and for virgin resin samples are shown in Table 5.3 and 5.4 for pH 6.0 and 6.8 tests, respectively. Three of the four resin columns (all but the SIR-700 pH 6.8 column) were divided into thirds. For example, the "top third" refers to the resin first in contact with influent water and so forth. The SIR-700 resin run at an influent pH of 6.8 was only divided into halves when taken offline early on in pilot testing; later, it was decided to obtain more resolution in the XRF testing, so the last three columns were divided into thirds rather than halves.

As shown in the tables, significant amounts of chromium, vanadium, copper, and uranium were accumulated on the resins. Duolite A7 added a significant amount of sulfur, whereas SIR-700 resin appeared to leach sulfur from the resin. Phosphorous (likely as phosphate) was accumulated to a slight degree on the resins.

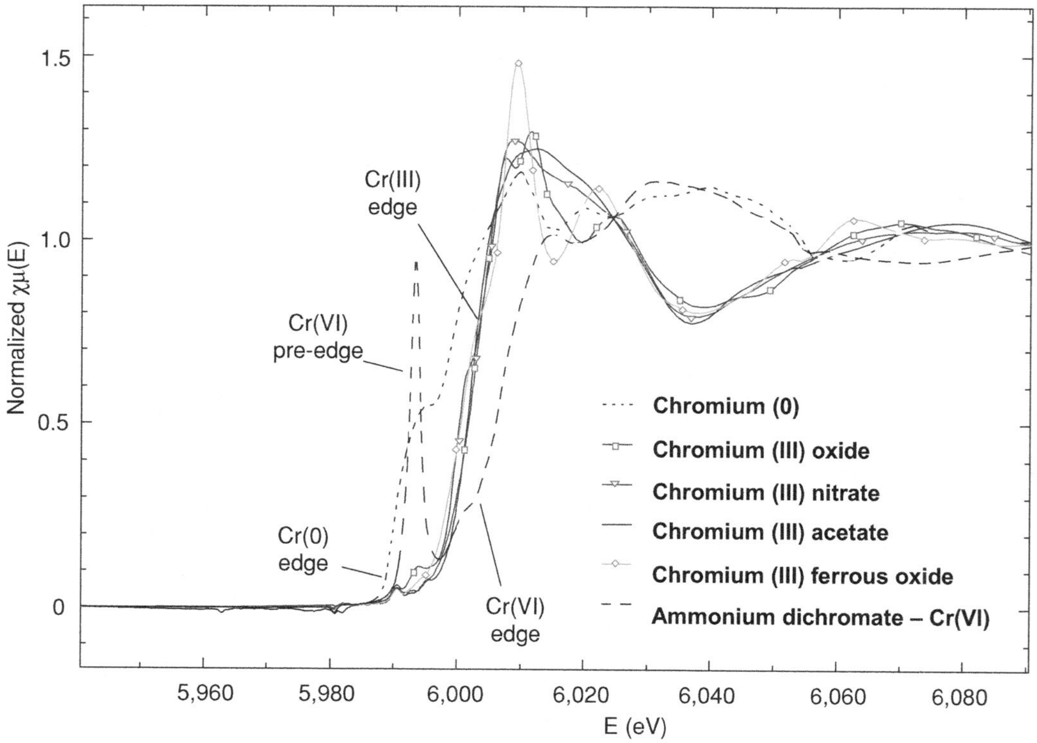

Figure 5.3 Cr K-edge XANES spectra of known Cr reference compounds

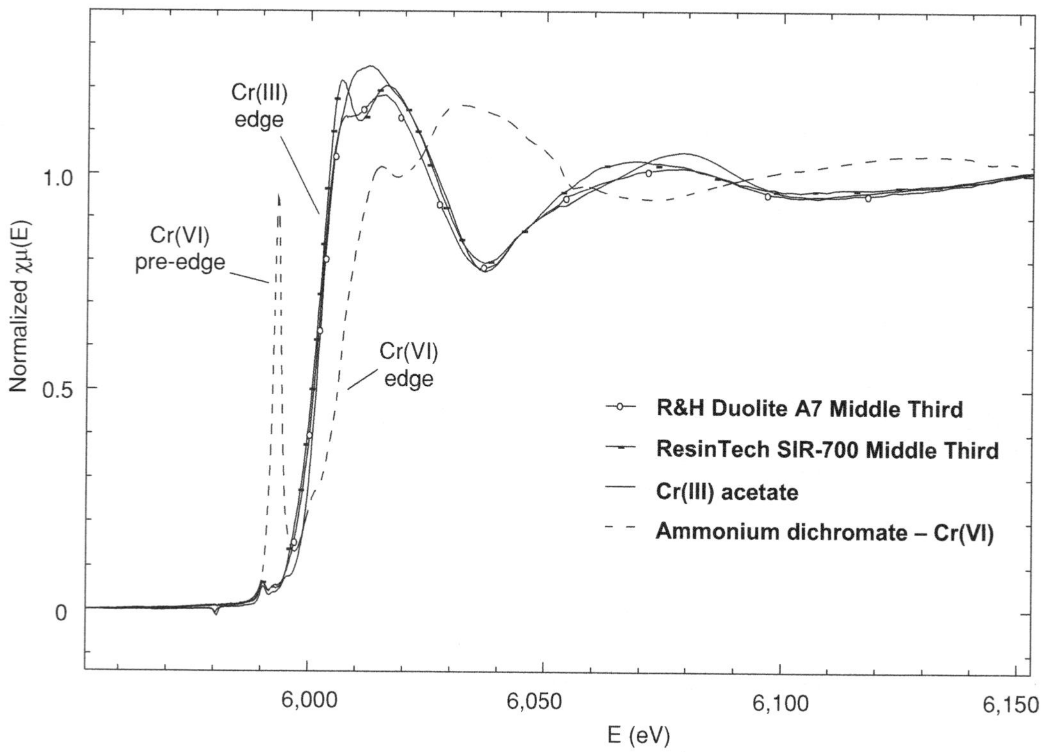

Figure 5.4 Cr K-edge XANES spectra of resin samples and selected Cr reference compounds

Table 5.3
XRF results of virgin and spent resin samples (pH 6.0)

	Duolite A7				SIR-700			
Element	Virgin resin	Top third	Middle third	Bottom third	Virgin resin	Top third	Middle third	Bottom third
Aluminum	164	165	68	330	<75	<61	<62	<63
Silicon	618	567	217	914	<40	241	214	97
Phosphorus	24	219	131	113	333	646	593	489
Sulfur	<2	6,348	6,176	7,214	64,370	37,970	40,240	41,450
Chlorine	31	3,916	3,705	4,044	3,656	4,644	4,779	4,780
Potassium	<9	589	283	148	<9	187	62	<10
Calcium	62	194	129	170	20	87	104	97
Vanadium	<4	1,330	1,915	2,440	<6	3,009	1,580	309
Chromium	8	14,600	10,450	7,583	<7	7,560	4,426	2,701
Iron	35	249	<8	31	39	85	95	<6
Cobalt	19	38	9	9	14	13	21	8
Copper	7	28,850	14,020	6,413	3.5	5,112	4,775	4,642
Zinc	3.3	<4	<3	15	1	<2	<2	2
Arsenic	<1	<1	<1	<1	<1	2	2	2
Bromine	0.3	136	105	92	3	99	98	89
Molybdenum	<12	64	50	30	<10	58	39	43
Iodine	<9	50	50	32	<9	43	37	48
Uranium	<1	1,885	860	410	0.7	781	209	35

NOTES: 1. Concentrations in the table are expressed in μg/g (ppm) dry resin.
2. At pH 6.0, Duolite A7 was run to 113,594 BV and SIR-700 was run to 113,566 BV.

Figure 5.5 shows the chromium concentration graphically as a function of resin position in the column and pH. As seen in the figure, Duolite A7 resin receiving water at pH 6.8 appeared to be saturated under those conditions (i.e., chromium concentrations in the top were similar to those in the bottom of the column). In contrast, Duolite A7 resin treating pH 6.0 water showed higher accumulation at the top of the resin compared to the middle and bottom. As noted in Figure 4.4, Cr(VI) in the column effluent was approximately 15 μg/L, or almost at 50% breakthrough.

SIR-700 resin also showed a similar pattern of increased accumulation on the resin at lower pH. As observed in Figure 5.6, the SIR-700 resin did not retain as much chromium as the Duolite A7. However, at pH 6.0 the SIR-700 resin also exhibited higher accumulation at the top compared to the bottom, indicating that the resin had not reached full capacity. Leakage of chromium occurred from both resins before saturation was achieved.

As observed in Table 5.3, high chromium concentrations were observed along with high copper concentrations on both resins. Figure 5.7 shows a correlation plot of chromium and

Table 5.4
XRF results of virgin and spent resin samples (pH 6.8)

	Duolite A7				SIR-700		
Element	Virgin resin	Top third	Middle third	Bottom third	Virgin resin	Top half	Bottom half
Aluminum	164	69	123	95	95	<61	<60
Silicon	618	503	485	365	<40	167	207
Phosphorus	24	70	47	33	333	253	190
Sulfur	<2	2,473	1,749	1,460	64,370	34,530	34,810
Chlorine	31	1,773	1,244	1,050	3,656	3,182	3,018
Potassium	<9	147	122	142	<9	43	<9
Calcium	62	195	202	202	20	217	179
Vanadium	<4	1,002	1,381	1,485	<6	389	<5
Chromium	8	5,810	5,872	5,427	<7	631	399
Iron	35	233	<8	<6	39	60	56
Cobalt	19	19	10	6	14	11	17
Copper	7	34,230	11,310	3,441	4	1,858	152
Zinc	3	<5	38	69	1	8	9
Arsenic	<0.3	<0.7	<0.5	<0.4	<0.4	<0.4	<0.4
Bromine	0.3	77	50	47	3	47	28
Molybdenum	<12	<18	14	<12	<10	<11	<11
Iodine	<8	31	28	14	<9	13	<9
Uranium	<1	519	424	367	0.7	121	0.8

Notes: 1. Concentrations in the table are expressed in μg/g (ppm) dry resin.
2. At pH 6.8, Duolite A7 was run to 98,538 BV and SIR-700 was run to 7,731 BV.

copper, which indicates that the two elements are highly correlated for both the SIR-700 and Duolite A7 resins. The influent water copper concentration in the GS-3 well water was lower than Cr(VI) at approximately 20 μg/L through pilot-scale testing. Figure 5.7 demonstrates that the Duolite A7 had a higher affinity for copper than chromium. By contrast, SIR-700 showed a different trend of increasing chromium retention for a similar amount of copper. The mechanism of copper removal by the WBA resins is unknown at this time, but removals of the two different elements may offer clues about the chromium removal mechanisms by the two resins.

XRD Results

XRD, a geochemical analysis used to identify and characterize unknown crystalline materials, was also performed on the spent resins. The purpose of using XRD in the study was to determine if any crystalline precipitates [e.g., $Cr(OH)_3 \cdot 3H_2O$ or Cr_2O_3] could be detected on the spent

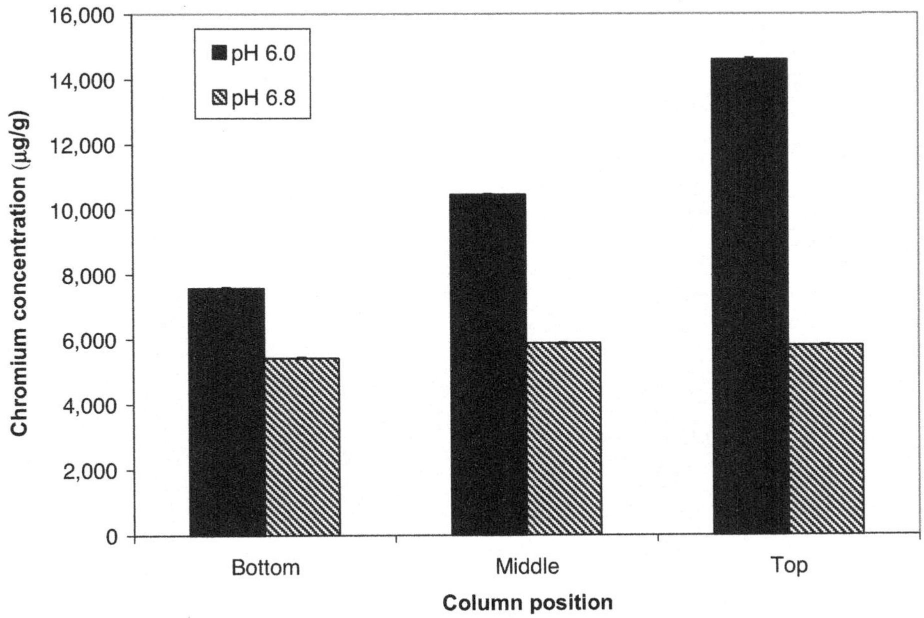

Figure 5.5 XRF results of Duolite A7 resin showing total Cr distribution in the resin columns as a function of depth and pH

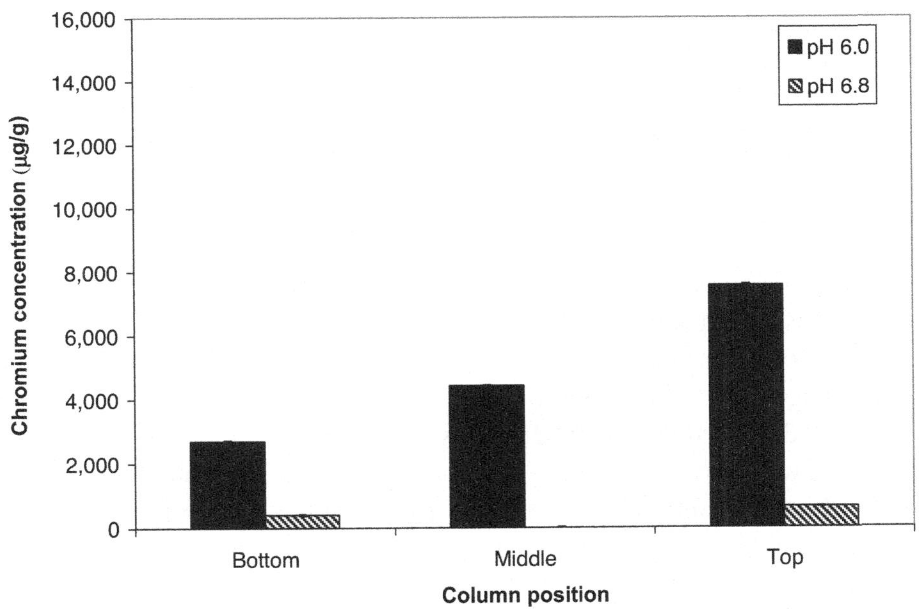

Figure 5.6 XRF results of SIR-700 resin showing total Cr distribution in the resin columns as a function of depth and pH (note pH 6.8 resin was divided into two sections)

resins. Figures 5.8 and 5.9 show the XRD patterns of spent Duolite A7 and SIR-700 resins, respectively. If crystalline solids were present at concentrations of approximately 1% or more, repeated sharp peaks would be present on the XRD patterns due to the diffraction of repeated crystal lattice structures, whereas amorphous materials (e.g., glass and liquids) produce a broad and continuous signal. As illustrated in Figures 5.8 and 5.9, no crystalline precipitates were found on either spent resin sample as evidenced by the lack of sharp peaks. These findings suggested

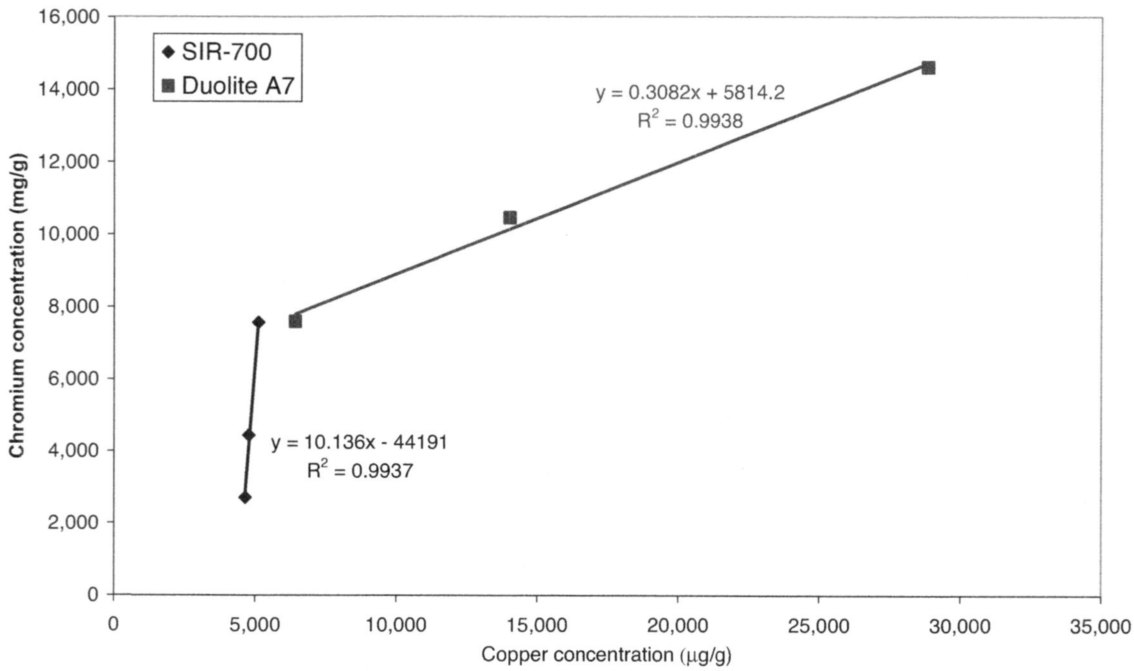

Figure 5.7 Correlation plot of chromium versus copper for the resins tested at pH 6.0

that chromium was not present as a crystalline Cr(III) compound such as $Cr(OH)_3{\cdot}3H_2O$ and Cr_2O_3 for the Duolite A7, which had total chromium concentrations exceeding 1% by weight. The single sharp peak observed on the SIR-700 virgin resin XRD pattern was believed to be caused by contamination during sample preparation.

SEM Results

SEM imaging was used to determine if chromium retained by the resin was in the form of precipitates on the resin surfaces or a more homogeneous distribution. The images shown in Figures 5.10 and 5.11 compare virgin Duolite A7 resin with spent Duolite A7 (pH 6.0, top third of the column). The images were taken without changing brightness and contrast settings. Two key observations were made based on the SEM analyses: (1) the spent resin generally had a higher brightness compared with the virgin resin due the increase in the abundance of metals (mainly from Cu and Cr and maybe some U), and (2) the higher degree of brightness was homogenously distributed throughout the resin and not present as visible chromium precipitates, which would be observed as "hot spots" in the SEM imaging.

Mechanism of Cr(VI) Removal by WBA Resins

In the pilot-scale column evaluation, the Duolite A7 resin once again demonstrated exceptional Cr(VI) removal capacity. Another WBA resin, SIR-700, also showed a high Cr(VI) capacity. XANES analysis indicated that the chromium was present as Cr(III). However, the specific mechanism causing the conversion of Cr(VI) to Cr(III) is not known. The importance of understanding the mechanism of Cr(VI) removal by the WBA resins lies in the ability to

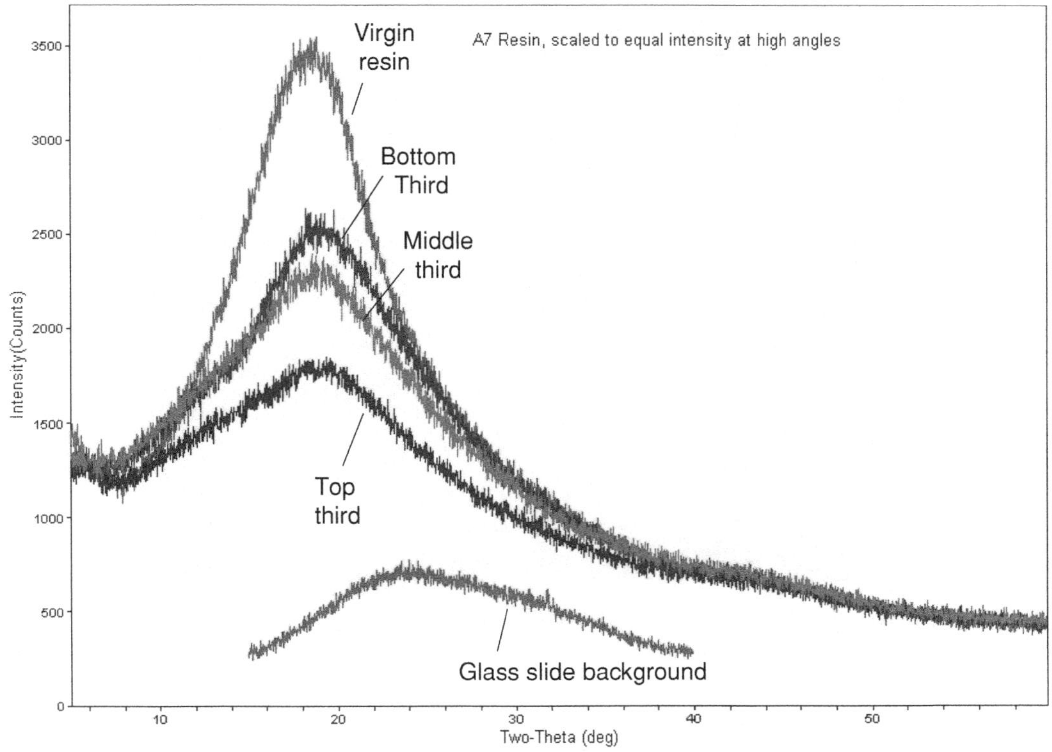

Figure 5.8 XRD patterns of spent Duolite A7 resin (pH 6.0)

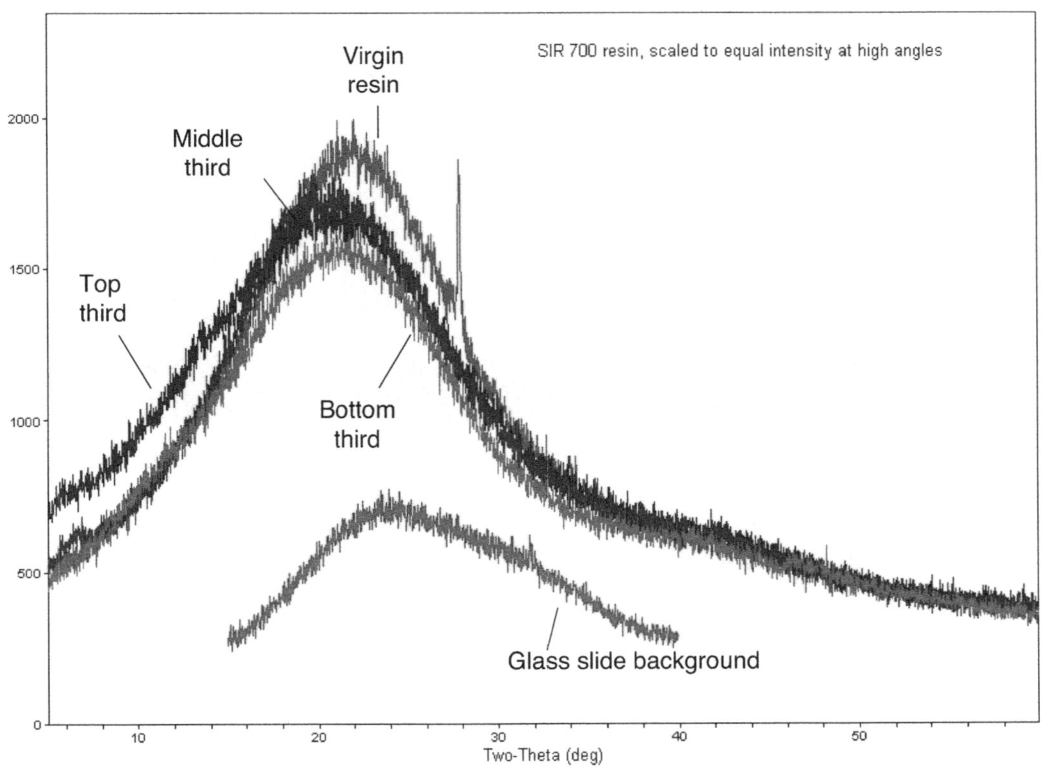

Figure 5.9 XRD patterns of spent SIR-700 resin (pH 6.0)

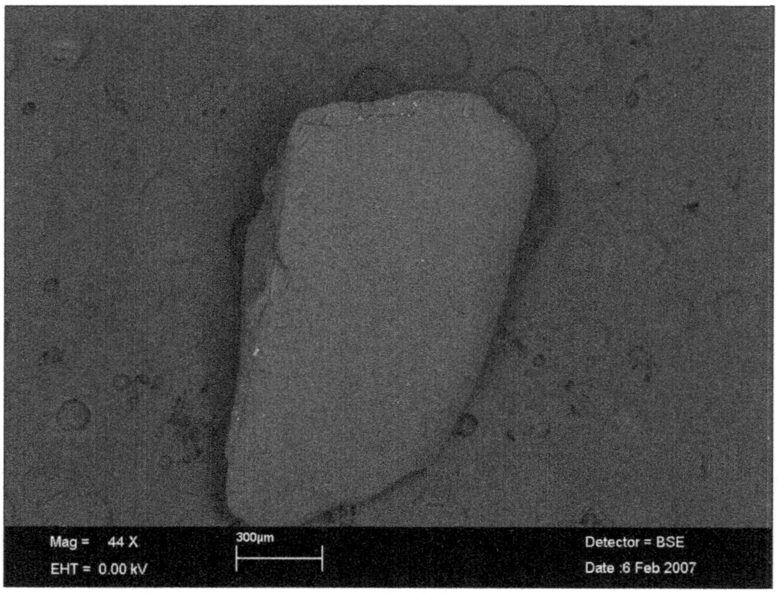

Figure 5.10 SEM image of virgin Duolite A7 resin – 44× magnification

Figure 5.11 SEM image of spent Duolite A7 resin (pH 6.0, top third) – 44× magnification

understand factors affecting removal performance and to reduce the chance of unexpected operational problems at full scale.

The lower operating pH (pH = 6.0) was clearly one factor contributing the improved Cr(VI) removal performance by the WBA resins. As illustrated in the aqueous speciation diagram below (Figure 5.12), the dominant Cr(VI) species is $HCrO_4^-$ at acidic pH (less than 6.5) while CrO_4^{2-} is dominant at alkaline pH (Sengupta and Clifford 1986). As $HCrO_4^-$ occupies only half the number of ion exchange sites per chromium atom as compared to CrO_4^{2-}, the lower operating pH is favorable for chromium ion exchange. In addition, the protonation of amine functional

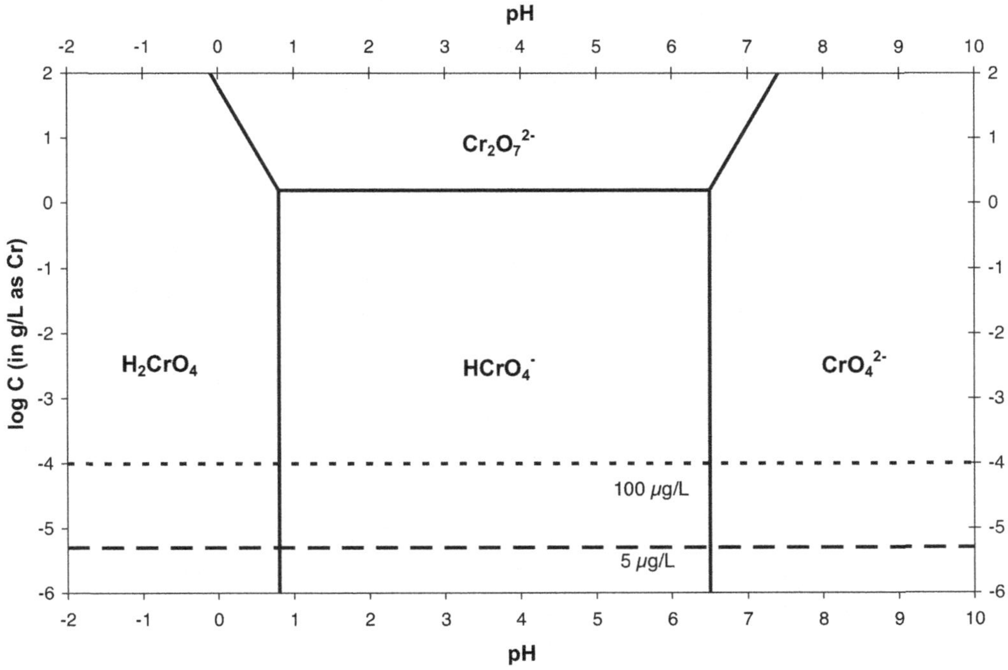

Note: The dashed lines in the diagram indicate the range of Cr(VI) concentration (5 to 100 μg/L) that is of interest in the Glendale study.

Figure 5.12 Aqueous speciation diagram showing the relative distribution of Cr(VI) species in water as a function of pH and total Cr concentration

groups on the WBA resins at pH values less than 6.5 enables the positively charged exchange sites to attract $HCrO_4^-$. Further, the competition of hydroxyl ions with chromate ions for WBA exchange sites is also less at acidic pH (Clifford 1999).

However, pH alone can not explain the extraordinary Cr(VI) removal capacity by the WBA resins as compared to conventional SBA exchange resins. The residual analyses of spent Duolite A7 and SIR-700 resins using XANES spectroscopy confirmed that a redox mechanism was involved in Cr(VI) removal. In spite of the differences in structures and functional groups between Duolite A7 and SIR-700 resins, a general Cr(VI) removal mechanism by these two WBA resins was hypothesized to include the following steps: (1) most Cr(VI) species in the aqueous phase is present as $HCrO_4^-$ at pH 6.0; (2) $HCrO_4^-$ is exchanged onto the functional groups of the WBA resins; (3) $HCrO_4^-$ is then reduced to Cr(III) by an unidentified electron donor (speculated to be either the function groups or the backbone of the resin material), which could be a rate limiting step; and (4) the Cr(III) species ultimately precipitates on the resin surface as amorphous $Cr(OH)_3$ or forms complexes with resin moieties. More research is needed to understand the complete mechanism behind the high Cr(VI) removal capacity by the WBA resins.

SUMMARY

The findings from the resin residuals analyses conducted in this study are summarized below:

- The spent Duolite A7 and SIR-700 resins (operated at pH 6.0) both passed the federal TCLP test but failed the California WET evaluation based on total chromium concentrations. Therefore, the resin residuals would be classified as hazardous waste for disposal in California.
- Since total uranium concentration on the spent Duolite A7 resin exceeded 500 μg/g, the resin could be characterized as low-level mixed radioactive waste and subject to radioactive waste disposal regulations. The operating life of the Duolite A7 resin should be limited to avoid complex radioactive waste disposal issues. For other utilities with uranium concentrations lower than approximately 1.4 pCi/L, uranium accumulation on WBA resins may not be a concern.
- XANES spectroscopy showed that the dominant chromium species retained on both resins was Cr(III), indicating that Cr(VI) was reduced to Cr(III) during the resin operation. Note that no Cr(III) leakage through the columns were observed at a constant pH (unlike in the Phase II Pilot-Scale Study where pH fluctuated).
- As observed by XRF, significant amounts of chromium, sulfur, vanadium, and copper accumulated on both spent resins.
- No crystalline precipitates were identified in the spent resins using XRD.
- Using SEM imaging, no precipitates were observed on the resin, indicating a fairly homogeneous distribution of chromium on the resin.
- The mechanism(s) underlying the high Cr(VI) removal capacity of WBA resins are still not completely understood, but testing revealed that reduction of Cr(VI) to Cr(III) is part of the process and a precipitated, crystalline Cr(III) compound is unlikely.

CHAPTER 6
COST ESTIMATES

As part of the Glendale Phase II Pilot Study, cost estimates were developed for the effective Cr(VI) removal systems (MEC 2005). These cost estimates were considered "preliminary" at that point due to information gaps in the WBA resin process. The cost information was thus updated as part of the Phase III Bridge Project using (1) additional information from the investigation of WBA resins (including residuals characterization); (2) consideration of ancillary facilities to the treatment process itself, which might include pre-filtration, on-site regulatory storage, repumping, and miscellaneous yard improvements, which can be an integral component of overall treatment; and (3) significant increases in material costs that have occurred since the development of the Phase II cost estimate.

Cost estimates were updated for three effective Cr(VI) removal technologies: weak-base anion exchange, strong-base anion exchange, and reduction/coagulation/filtration. Estimates for all three are presented in this chapter along with the conceptual drawings of each treatment process.

COST DEVELOPMENT APPROACH

Technology cost estimates were refined by obtaining updated quotes for installation of commercially-available SBA and WBA columns directly from vendors of those technologies (e.g., BasinWater 2006, Siemens 2006a, Siemens 2006b), requesting technology cost information from several vendors for the RCF process, and developing independent cost information using existing literature, professional judgment, and industry cost models. Estimates were developed for two system sizes with design flow rates of 500 gpm and 1,000 gpm representing single well treatment at GS-3 and joint treatment of two GN wells, respectively.

The American Association of Cost Engineers (AACE) defines five categories of estimates in an effort to establish expected accuracy range for various types of cost estimates (Table 6.1). The objective of this work was to refine initial conceptual screening cost estimates to provide study or feasibility-level costs. It is expected that an estimate of this type would be a Class 4 estimate accurate within –15% or +30%. Basically, the estimates developed in this project were based on process flow diagrams developed from pilot testing experience rather than conceptual-level technology screening, indicating that the estimates would be at least Class 4 rather than Class 5 estimates. However, many tasks in preliminary design had not yet been initiated at this point, thus making the costs Class 4 rather than Class 3 estimates. Additional details on the selection of the estimate class are provided in the AACE International Recommended Practice document (AACE, 2000).

Vendor Cost Solicitation

Technology cost information was requested from vendors for those technologies that were most promising in Phase II pilot tests (McGuire et al. 2006). This included Siemens in partnership with Rohm and Haas (Siemens/R&H) weak- and strong-base ion exchange. In addition, we requested a cost quotation from BasinWater, a strong-base ion exchange vendor with a containerized treatment system. Finally, various vendors were contacted for quotes on unit processes within

Table 6.1
AACE cost estimation classification

Estimate class	Level of project definition (expressed as % of complete definition)	End usage – typical purpose of estimate	Typical budget estimate accuracy
Class 5	0 to 2%	Concept screening	–30% to +50%
Class 4	1 to 15%	Study or feasibility	–15% to +30%
Class 3	10 to 40%	Budget, authorization, or control	–15% to +30%
Class 2	30 to 70%	Control or bid/tender	–5% to +15%
Class 1	50 to 100%	Check estimate on bid/tender	–5% to +15%

the RCF treatment process. Vendor-supplied cost quotations were amended using independently-developed capital costs and operations and maintenance (O&M) estimates, as discussed below.

Independent Cost Development

Capital cost estimates were developed using the MasterFormat™ framework established by the Construction Specifications Institute. MasterFormat™ provides an organizational structure for performance-based construction specifications and costs. The Malcolm Pirnie Standard Specifications Format conforms closely to the Construction Specification Institute's MasterFormat 1995. The Malcolm Pirnie Standard Specifications Format is based on a Division-Section concept, in which each division is identified by a division number and title. The division title is a broad generic heading based on an interrelationship of place, trade, function, or material. The eighteen divisions that are constant in sequence, name, and number, are listed below:

Division 0 - Bidding Requirements, Contract Forms, Conditions of the Contract
Division 1 - General Requirements
Division 2 - Site Construction
Division 3 - Concrete
Division 4 - Masonry
Division 5 - Metals
Division 6 - Wood and Plastics
Division 7 - Thermal and Moisture Protection
Division 8 - Doors and Windows
Division 9 - Finishes
Division 10 - Specialties
Division 11 - Equipment
Division 12 - Furnishings
Division 13 - Special Construction
Division 14 - Conveyance Systems
Division 15 - Mechanical
Division 16 - Electrical
Division 17 - Instrumentation and Controls

Table 6.2
Standard capital cost multipliers

Capital cost	Cost multiplier
Insurance	2.5%
Bonds	2.0%
Contractor overhead and profit	10%
Engineering	10%
Contingency	20%

The division titles are primarily an organizational device. Each division contains a group of related sections, with each section generally constituting a unit of work or a single entity such as a particular material, product, or item of equipment.

In addition to specific estimates associated with each division, several standard capital cost multipliers were applied to the total capital cost (Table 6.2).

General Cost Assumptions

A discount rate of 5% with a recovery period of 20 years was used to annualize all technology capital costs. All costs are expressed in September 2006 dollars. Labor costs were estimated for each technology based on full-time equivalents (FTEs) with a loaded annual salary of $100,000 (typical estimate for Glendale, California).

TECHNOLOGY COST ESTIMATES

Weak-Base Anion Exchange

Treatment Process Description

A WBA exchange system would consist of lead/lag resin vessels with upstream acid addition. Due to its high capacity and difficulty in regeneration, WBA resin is intended be used as a once-through, non-regenerable media. Figure 6.1 provides a process flow schematic of a WBA system. Note that the water from the well to the treatment system will be kept under pressure and an aeration step will not be necessary (i.e., similar to the pilot study configuration rather than the mini-column evaluation).

Acid requirements for pH depression to 6.0 determined from the Phase III Bridge Project were approximately 0.00023 gallons of 31% HCl per gallon of water treated. Siemens/R&H specified a volumetric design flow rate for Duolite A7 media of approximately 2.5 gpm/ft^3 (corresponding to a service flow rate of 10 gpm/ft^2), bed volumes of 200 cubic feet, and 8-ft. diameter vessels. Cost estimates were based on removal of Cr(VI) to approximately 100,000 bed volumes, which corresponds to approximately 207 days of operation before resin is replaced. Siemens/R&H estimated that a weekly low-volume backwash may be necessary to reclassify the media bed, which would yield non-hazardous backwash water.

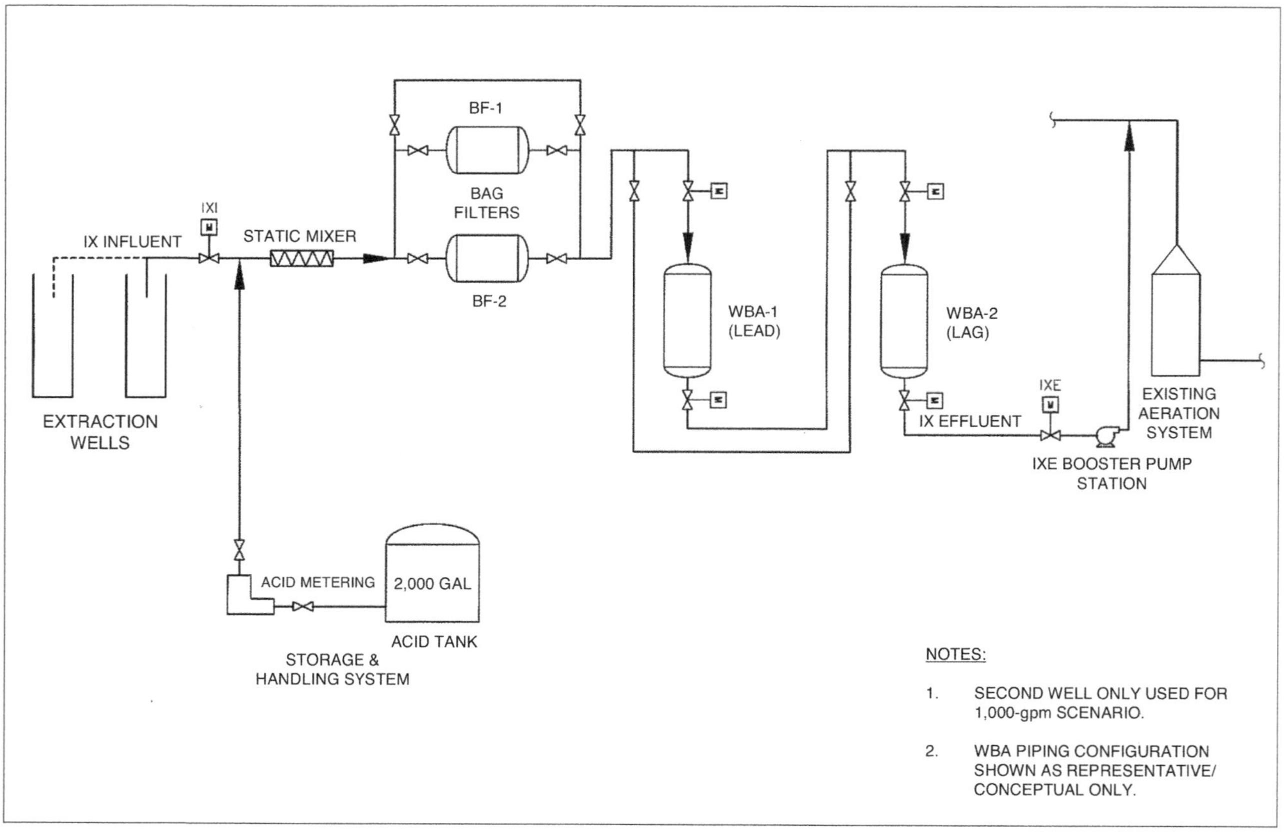

Figure 6.1 Process flow schematic of a WBA system

Capital Cost Development

For the 500 gpm demonstration system, two 8-ft. diameter vessels would be plumbed in a lead/lag configuration. By comparison, a 1,000 gpm demonstration system would have two 12-ft. diameter vessels. Other equipment included in this cost estimate are bag filters, an HCl storage and handling system, a 16,000-gallon liquid waste equalization tank, and a centrifugal pump for regulating the spent backwash water flow to the sewer. Two parallel bag filter housings (5 micron filters inside 3-ft. diameter housings) were considered in the estimate. For acid feed, cost estimates were developed using metering pumps capable of 288 gallons per day, a 2,000-gallon HCl storage tank with a 3,000-gallon secondary containment fiberglass reinforced plastic (FRP) tank, and a scrubber system for acid vapor during acid offloading.

The initial resin inventory (400 cubic feet – divided into two vessels) was also included in the capital cost estimate at a rate of $500/cubic foot (i.e., the estimate provided by Siemens/R&H, which includes loading and disposal of spent resin). Additional resin loads after the first fill were accounted for in the O&M costs. Other one-time capital costs are shown by division in Appendix B.

Operation and Maintenance Cost Development

Operating costs for the WBA system were provided by Siemens/R&H and adjusted as described earlier. Estimates included the following:

- Media replacement every 207 days, at $500 per cubic foot,
- Hydrochloric acid for pH depression, at $1.15/gallon for 2,000 gallon loads of 31% HCl (quote obtained from Basic Chemical, October 2006),
- Non-hazardous liquid waste disposal for backwash, at $3.22 per 1,000 gallons,
- Bag filter replacement costs, assuming monthly replacement of filters,
- Effluent booster pump energy costs, and
- Labor costs for 0.125 FTE.

The Siemens/R&H WBA system costs are dominated by O&M costs comprised largely of resin and acid costs. The resin replacement costs are driven by two key assumptions: (1) the anticipated resin usage rate is based on 100,000 bed volumes to breakthrough; and (2) the cost of resin estimated at $500/cf. The cost of WBA resins has varied significantly in the past two years, increasing from $350/cf to $500/cf.

Since the WBA resin requires pH depression to approximately 6.0, the acid needs are substantial in Glendale's groundwater, which is significantly buffered with a high alkalinity. As discussed in Chapter 3, Phase III Bridge Project testing indicated that the pressurized GS-3 well water at GS-3 has a pH of 6.8 rather than the pH of 7.2–7.3 that was routinely reported by the laboratory. To achieve a pH of 6.0, approximately 0.00023 gallons of 31% HCl will be needed, which corresponds to about 165 gallons of 31% HCl per day for a well pumping rate of 500 gpm. Although not tested at pilot scale, cost savings may be realized if sulfuric acid can be used; this may be a variable that could be tested in the demonstration study.

Residuals streams from the WBA system are limited to spent resin and backwash water. Non-hazardous backwash water could be sent to either the Los Angeles sewer facilities (for the Southern wells) or to the Glendale sewer (for the Northern wells). Spent resin will be hazardous due to high chromium levels, as shown by the California WET test results in Chapter 5. Hazardous waste disposal costs for the WBA resins were estimated at $445 per ton and are included in the O&M estimates for the WBA system.

Estimated Cost Range

Based on vendor-provided cost estimates and an independent analysis of costs, estimates were developed for 500 and 1,000 gpm WBA systems. For WBA, a third case was considered – a retrofit of two GAC vessels at the GS-3 site for treating 500 gpm. This third option was estimated to save approximately $700,000 for capital costs associated with the WBA system.

Figure 6.2 shows the capital cost and annual O&M cost estimates for 500 and 1,000 gpm systems. Figure 6.3 displays the annualized costs in dollars per acre-foot of water treated.

As noted previously and shown in Figure 6.3, the O&M costs are the largest cost component of the annualized costs, which is due to both resin replacement frequency and acid needs. Since resin costs dramatically increased from 2005 estimates likely due to market pricing, the potential exists for resin costs to come down, particularly if other WBA resins are available to provide cost competition.

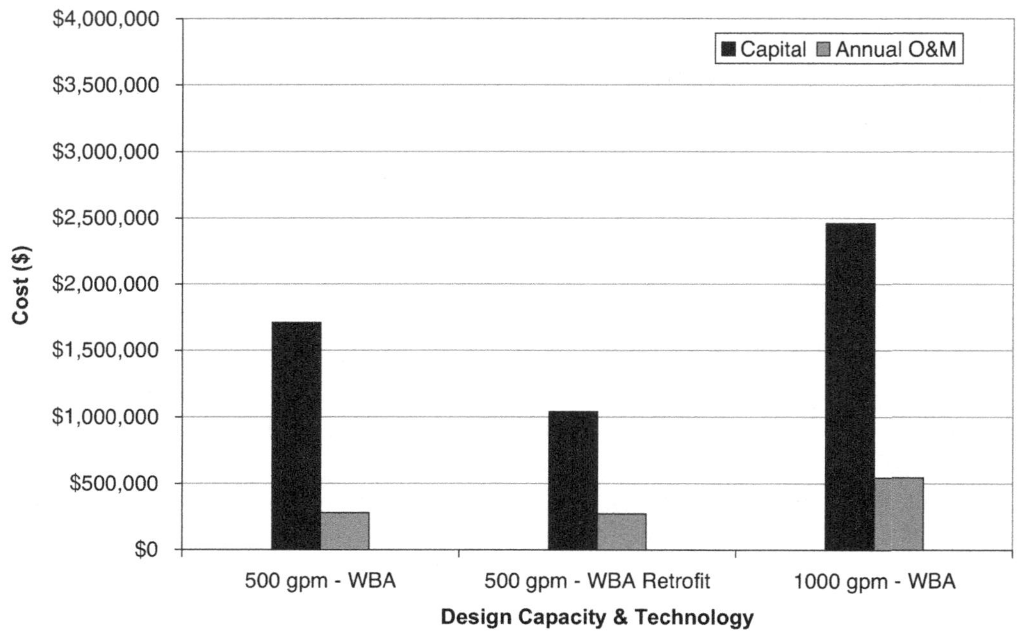

Figure 6.2 Capital cost estimates and annual O&M of the WBA system

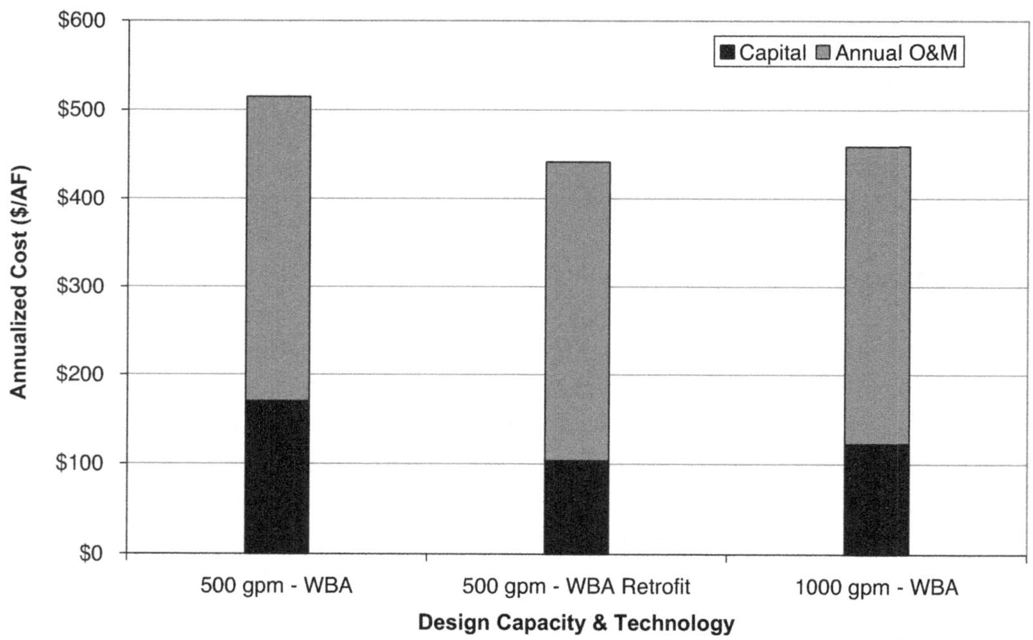

Figure 6.3 Annualized cost estimates of the WBA system

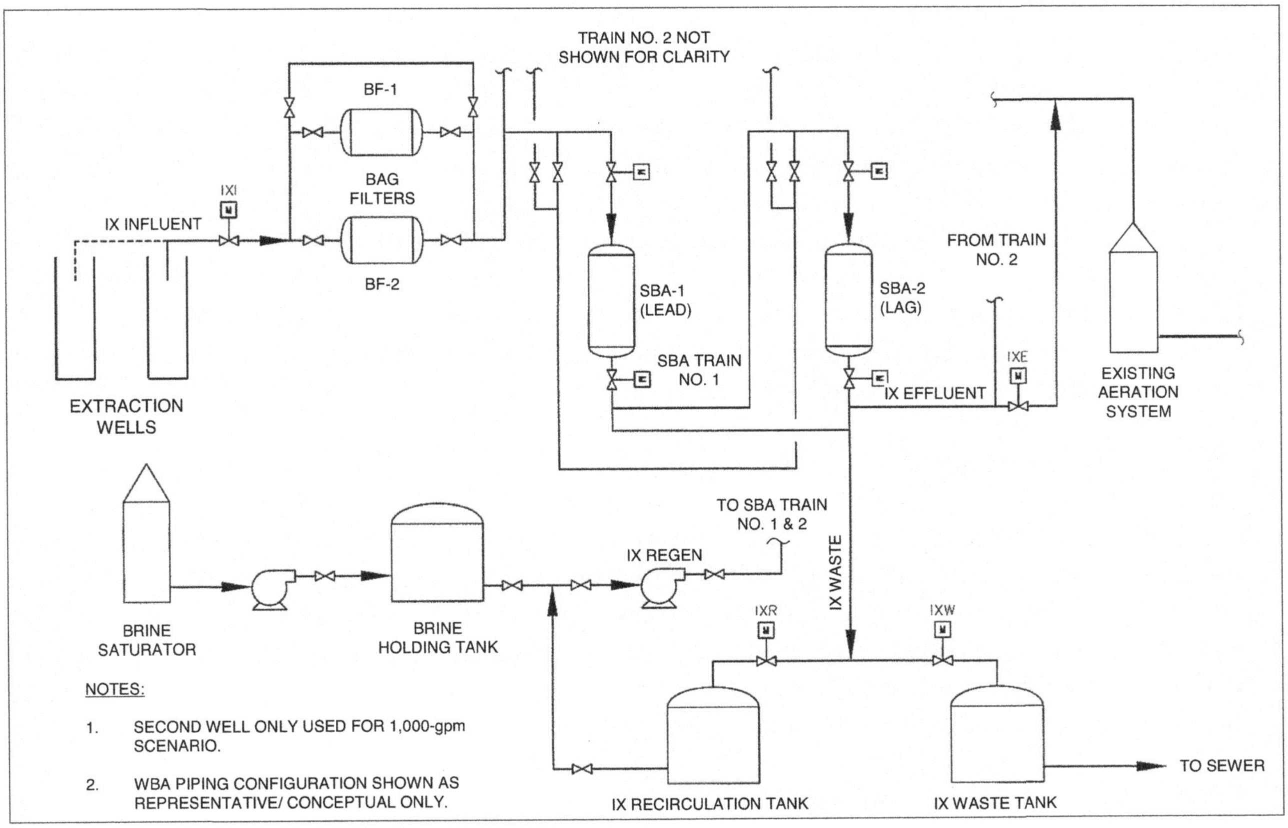

Figure 6.4 General process flow schematic of an SBA system

Strong-Base Anion Exchange

Treatment Process Description

In the Phase II cost estimate, Siemens/R&H provided a quotation for an SBA system. However, after obtaining experience with an installation in Colby, Kansas, Siemens no longer offers regenerable SBA for Cr(VI) removal (Peschman, 2006). During the Kansas testing, Siemens encountered difficulty in regenerating the resin to full capacity after approximately 12 to 15 regenerations. Consequently, Siemens now markets only the WBA system and a single-pass SBA system.

In contrast to Siemens' strategy, BasinWater provided a quotation for a regenerable SBA system with brine treatment. No previous estimate had been provided by BasinWater in the Phase II Pilot Study. The BasinWater system uses multiple beds in parallel for treatment, which enables regeneration of some beds while others are in service. For a 500 gpm system, seven beds (3-ft. diameter with 25 cf of resin) would treat the water at a given time and 3 would be in various stages of regeneration (brine regeneration and rinsing). In a 1,000 gpm system, 13 beds would be online and 3 would be in regeneration during a typical operation. The BasinWater system uses a Type I SBA resin from ResinTech (SBG1). A general process flow schematic illustrating the different components of a SBA system, including brine regeneration and treatment, is shown in Figure 6.4. Note that a process flow schematic specific to the BasinWater system was not available.

An important point about BasinWater's SBA system is their contention that the regeneration process is much more efficient than typical SBA systems. As a result, the amount of salt BasinWater purports to use will be much lower (59 lb/AF water treated) than other brine regeneration applications (such as Calgon's ISEP process for perchlorate, which uses about 1,600 to 2,400 lb/AF). Partial data were provided for a BasinWater chromate removal system in Stockton, California; however, no proof of effective regeneration after 12 to 15 cycles was provided.

The BasinWater system also contains a brine processing unit to process the spent brine, thus rendering it non-hazardous. In this system, Cr(VI) is reduced using ferrous salt and precipitated. Based on Phase II testing, the precipitated solid waste component will be hazardous by California WET standards due to chromium concentrations. Additional details on the brine processing unit have not been provided by BasinWater.

Capital Cost Development

Capital cost estimates provided by BasinWater include a treatment module (mobile container), 6,500-gallon salt storage (saturator) tank, two 6,500-gallon brine wastewater storage tanks, and a brine processing unit. BasinWater provided several options for their system, including purchase, a take-or-pay option with a 10-year service agreement, and a monthly standby with a water service agreement. For purposes of cost comparison, the purchase option was considered in this evaluation. Other one-time capital costs are shown by division in Appendix B.

Operation and Maintenance Cost Development

Operating costs for the SBA system were provided by BasinWater and adjusted as appropriate. Estimates included the following:

- Salt usage, estimated at 59 lb per acre-foot of water treated at a cost of $100/ton for salt
- Non-hazardous backwash water disposal, at $3.22 per 1,000 gallons
- Non-hazardous brine disposal, including trucking costs and discharge fees to Hyperion for a total cost of $0.15/gallon
- Solid hazardous waste disposal at a rate of $445/ton
- Bag filter replacement costs, assuming monthly filter replacement
- Effluent booster pump energy costs
- Labor costs for 0.5 FTE (note: more operator attention will be required compared to WBA due to brine treatment)

Residuals disposal cost estimates dominated the O&M costs due to the need for trucking non-hazardous (treated) brine to a sewer connection leading to the Los Angeles Hyperion Wastewater Treatment Plant. This step is required since the Los Angeles-Glendale Water Reclamation Plant will not accept brine. The potential for brine recycle has not been offered by BasinWater and may offer the potential for reduced spent brine volumes, as indicated by the Phase II Pilot-Scale Study in which SBA resin could be effectively regenerated using a saturated brine solution recycled up to 7 times. The volume of non-hazardous brine includes regenerant and slow rinse liquid waste. Other sources of residuals waste include fast rinse and backwash liquid waste, which will

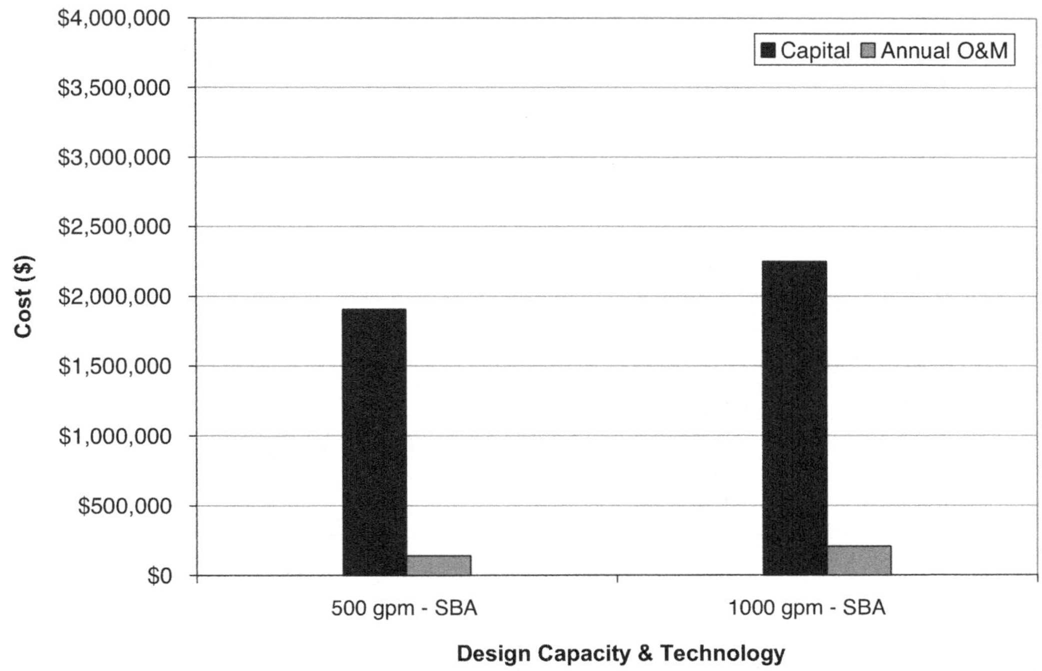

Figure 6.5 Capital cost estimates of the SBA treatment system

be sent to the sewer as non-hazardous liquid waste, and solid hazardous waste from brine processing.

Estimated Cost Range

Based on vendor-provided cost estimates and an independent analysis of costs, cost estimates were prepared for SBA treatment. Figure 6.5 shows the total capital and annual O&M cost estimates for 500 and 1,000 gpm systems. Figure 6.6 displays the annualized costs for SBA systems in dollars per acre-foot of water treated. As shown in Figure 6.6, the O&M and capital costs are similar on an annualized basis.

Several variables associated with the SBA technology are unknowns with the potential to significantly impact SBA treatment process feasibility, including:

- BasinWater has not released data proving the capabilities of the brine processing unit for rendering the liquid component of the brine non-hazardous. Limited research has demonstrated this brine treatment approach, and it is uncertain if the capital (e.g., dewatering equipment) and O&M (e.g., increased labor necessary to manage spent brine treatment and sludge disposal, estimated here to be an extra 0.5 FTE) estimates provided by BasinWater are accurate. Consequently, the residuals cost estimates for the brine treatment systems may not reflect all of the anticipated costs associated with this approach. At this time, however, these costs reflect information provided by BasinWater.
- The amount of salt required for BasinWater SBA regeneration is very low compared to regenerable SBA processes for perchlorate. If larger quantities are required than those

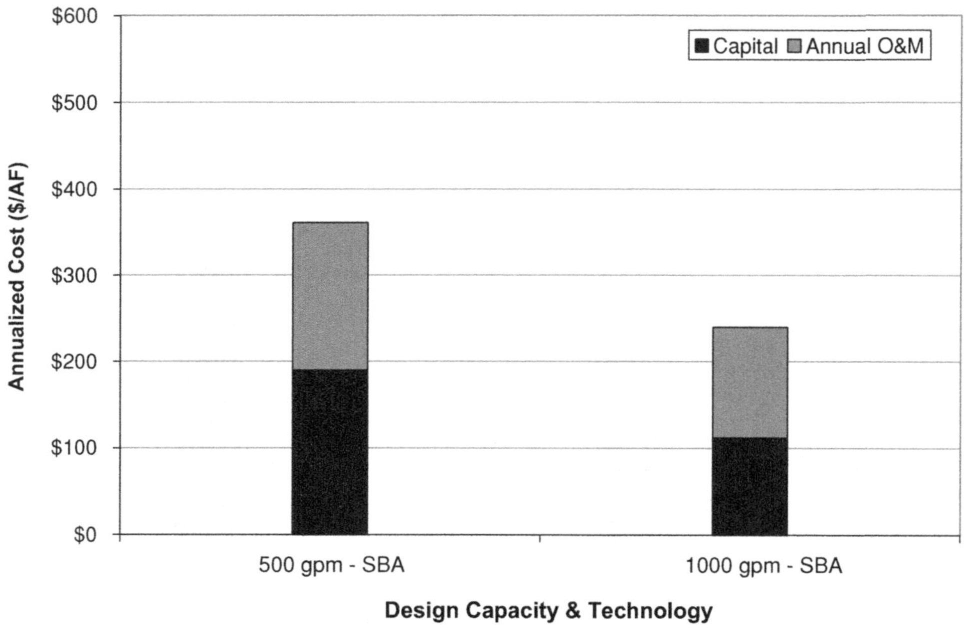

Figure 6.6 Annualized cost estimates of the SBA treatment system

estimated by BasinWater, the brine processing unit may be undersized and quantity of waste underestimated.

- Disposal of brine to a connection leading to the Hyperion sewer is not certain in the future. If disposal of brine to the Hyperion sewer system is not allowed at some point in the future, SBA treatment would not be feasible and/or cost-effective for Glendale. Installing SBA treatment at a Glendale facility would entail a high degree of risk.
- No data have been provided by BasinWater assuaging the concern that SBA resins may not be regenerable beyond 12 to 15 regenerations (as observed by Siemens).

Reduction/Coagulation/Filtration

Treatment Process Description

Reduction of Cr(VI) to Cr(III) with ferrous sulfate, subsequent coagulation with ferric iron, and then filtration was demonstrated to successfully remove Cr(VI) during pilot testing. The RCF pilot unit consisted of a reduction tank, aeration chamber, and dual-media granular filters. In addition, chemical feed points included those for pH control before the reduction tank and aeration columns, ferrous sulfate addition before the reduction tank, and filter aid polymer addition prior to the filters.

Since the RCF process had never been utilized to remove Cr(VI) from a potable drinking water supply, the RCF technology costs were estimated using a range of sources including quotes from vendors that could supply different unit processes. Figure 6.7 illustrates the RCF system process flow diagram.

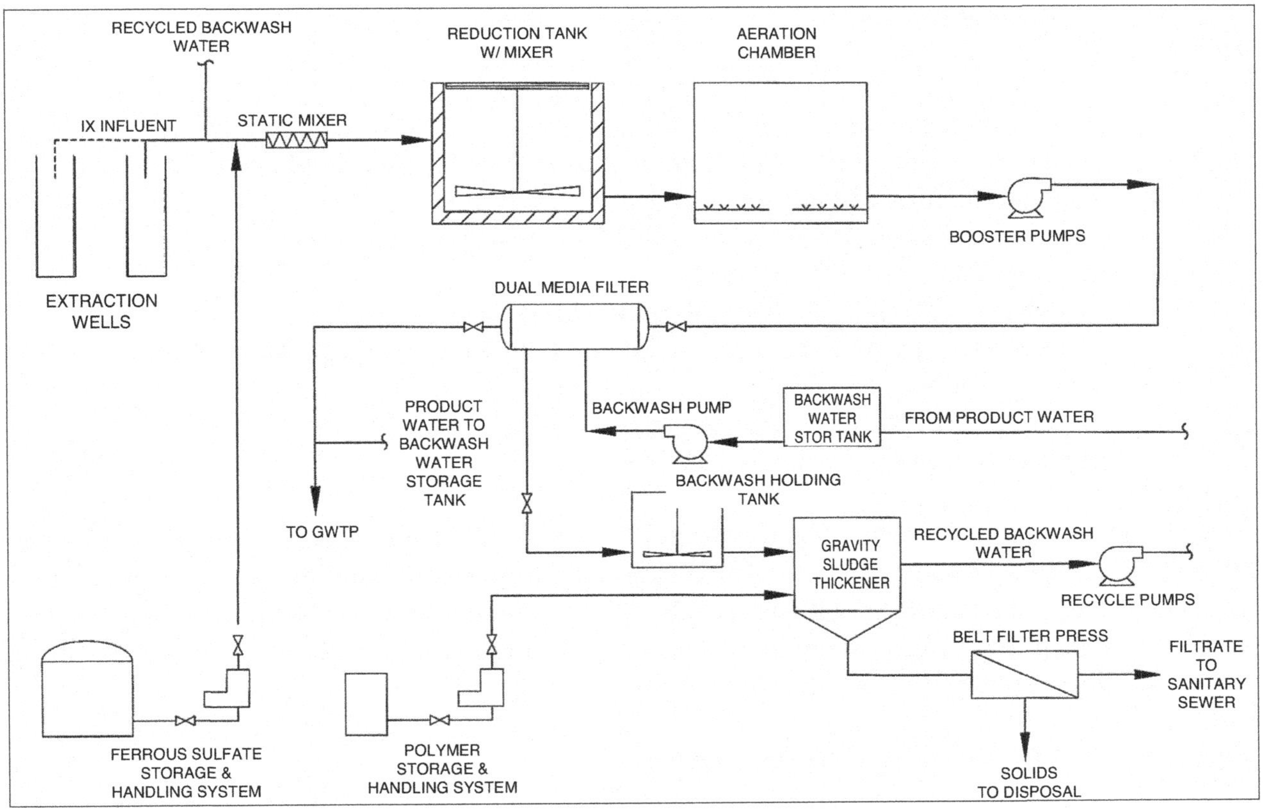

Figure 6.7 Process flow schematic of the RCF system

Pilot Testing Results

Pilot testing efforts identified optimal operating conditions for the RCF process. At ambient pH conditions and an Fe(II):Cr(VI) mass ratio of 25:1, the system continuously removed chromium [both Cr(VI) and Cr(III)] to below detectable levels for 48 hours before filter head loss and turbidity goals were exceeded. The pilot system was backwashed with air scouring and bed fluidization. The bed expansion rate was controlled at 20 to 30%, and the entire backwash procedure was complete within 10 to 12 minutes.

Waste minimization and disposal options for the chromium-containing backwash water solids were also investigated during pilot testing. It was determined that the backwash solids could be rapidly settled with low doses (0.2 to 1.0 mg/L) of high molecular weight polymer. The resultant settled backwash water may be suitable for recycle to the head of the system, while the settled backwash solids would be dewatered. Liquid waste from the dewatering process would be classified as non-hazardous and sent to the sewer. Dewatered solids would be classified and disposed of as hazardous solid waste.

Capital Cost Development

Capital costs for the RCF process include the chemical feed systems, reduction tank, aeration chamber, filters, backwash pumping, settler for backwash solids, and waste handling treatment equipment. Since the pilot testing identified ambient pH conditions as optimal, pH adjustment equipment has not been included in this estimate. One-time capital costs included the following:

- Reduction tank with mixers (30,000-gallon)
- Aeration chamber (10,000-gallon, with coarse air diffusers, and air compressor)
- Dual media filter (5-cell)
- Ferrous sulfate feed system, including chemical storage and pumps
- Polymer addition feed system
- Backwash water storage tank (30,000-gallon)
- Backwash water (post-backwash) holding tank with mixer (20,000-gallon)
- Gravity settler
- Belt filter press (1-meter)
- Booster pumps

The capital costs estimated for the RCF system are higher than the other technologies evaluated, primarily due to the cost of filters and backwash waste handling and treatment equipment required to dewater the backwash solids. As a potential alternative to granular media filters, microfiltration might offer cost savings. Alternate filtration strategies could be evaluated in the demonstration-scale preliminary design phase.

Operation and Maintenance Cost Development

Operational costs for the RCF process included the following:

- Ferrous sulfate costs, estimated at $0.60/gallon (5% Fe) based on a usage rate mass ratio of 25:1 Fe:Cr (2.5 mg/L as Fe for 100 μg/L Cr)
- Non-hazardous backwash water disposal, at $3.22 per 1,000 gallons
- Solid hazardous waste disposal at a rate of $445/ton
- Filter media replacement, assuming 10% media loss per year
- Effluent booster pump energy costs
- Labor costs for 0.5 FTE

Residuals streams produced by the RCF process include dewatering liquids classified as non-hazardous waste and sent to the sewer, as well as dewatered chromium-containing coagulation solids disposed of as hazardous solid waste. Residuals stream volumes were estimated based on full-scale operational conditions including 24-hour filter run times with a 15-minute backwash duration at 15 gpm/ft^2. These conditions resulted in a backwash water volume of 4% of the treated flow. A number of assumptions were then needed to calculate liquid and solid residuals components. The RCF residuals disposal cost estimates rely heavily on the operational assumptions for the process, including backwash duration and quantity, settled backwash water sludge volume, and dewatering efficiency. These operational assumptions dictate the total volumes of dewatering liquids classified as non-hazardous that are sent to the sewer, as well as dewatered chromium-containing coagulation solids disposed of as hazardous solid waste. With no available full-scale operational history for the RCF process, these assumptions were limited to industry experience and such texts as *Water Quality and Treatment* citing coagulation practice (Cornwell, 1999).

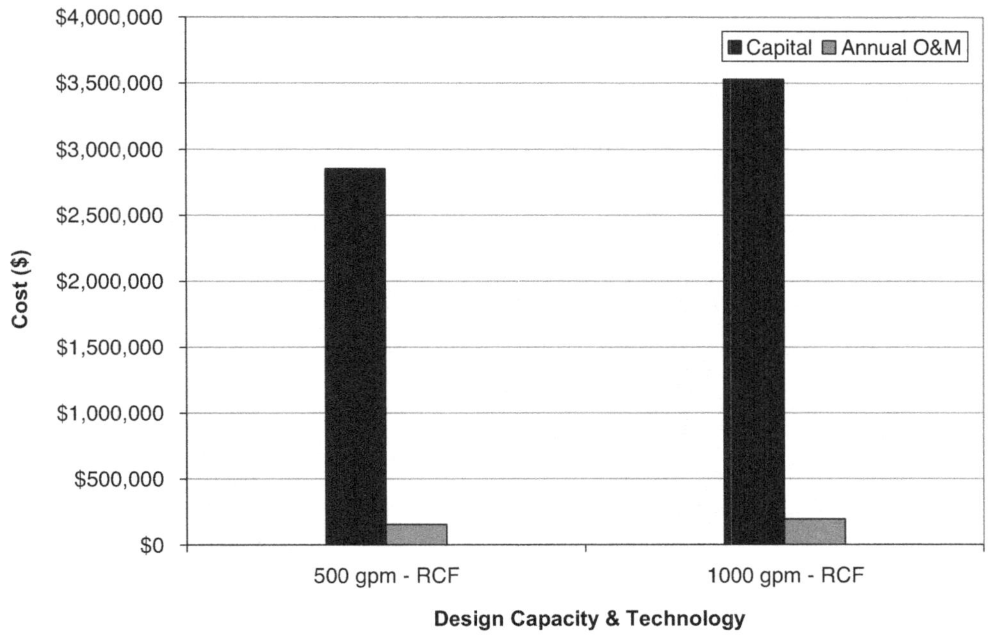

Figure 6.8 Capital cost estimates of the RCF treatment system

Of the backwash water, 0.58% was assumed to be settled sludge (based on calculations of ferrous doses), and the remaining 99.42% of the backwash water was assumed to be recycled to the head of the plant. Settled sludge was assumed to have 3% solids, and the filter press was assumed to have 80% dewatering efficiency (Cornwell, 1999). These assumptions led to an estimate of 28,000 gallons per day of non-hazardous sludge decant water sent to the sewer for a 500 gpm system – a small cost component. Solid residuals, on the other hand, were a significant cost at $445/ton and an estimated 20 tons produced per year.

Estimated Cost Range

Based on vendor-provided cost estimates of unit processes and an independent analysis, cost estimates were developed for the RCF treatment process for 500 and 1,000 gpm flows. Figure 6.8 shows total capital and annual O&M cost estimates for the two different size systems. Figure 6.9 displays the annualized costs for RCF systems in dollars per acre-foot of water treated. In general, the RCF system costs are characterized by relatively low O&M costs and high capital costs.

COST COMPARISON

Capital Cost

Figure 6.10 shows a comparison of the capital cost estimates for each of the three technologies at 500 and 1,000 gpm. For the WBA technology, the capital cost estimates were $1.0 million, $1.7 million, and $2.5 million for the retrofitted 500 gpm, new 500 gpm, and 1,000 gpm systems, respectively. For the SBA technology, the capital cost estimates were $1.9 million and $2.3 million for the 500 gpm and 1,000 gpm systems, respectively. For the RCF

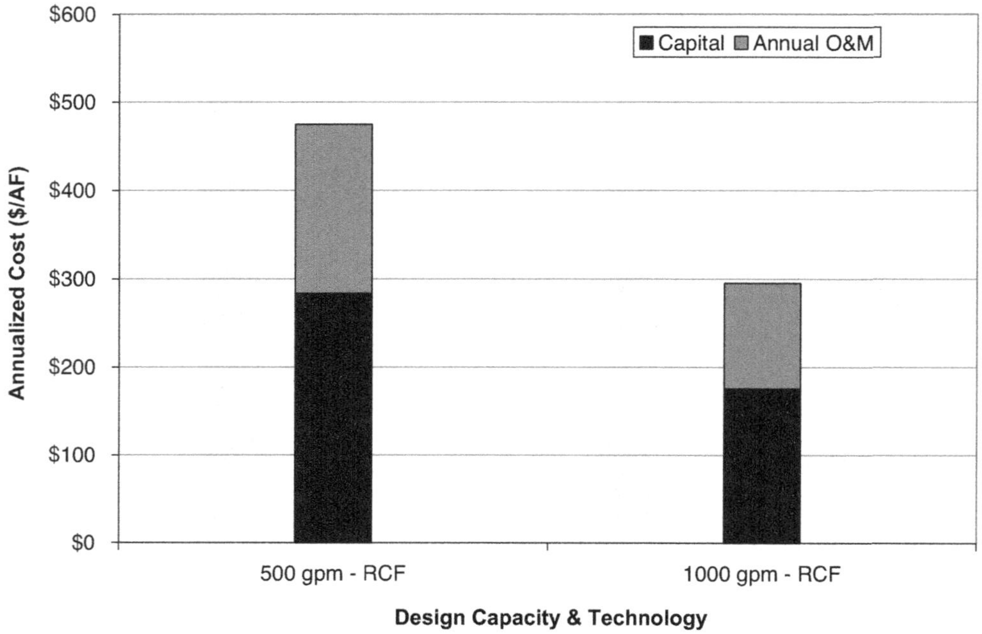

Figure 6.9 Annualized cost estimates of the RCF treatment system

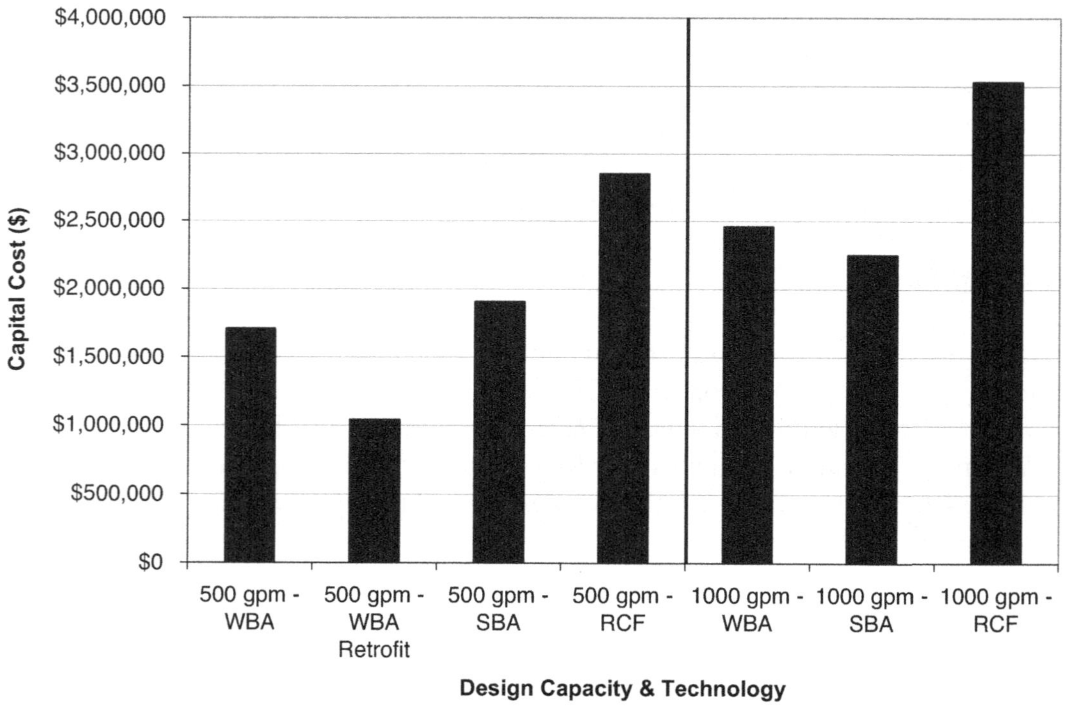

Figure 6.10 Comparison of capital cost estimates for the three technologies at two flow rates

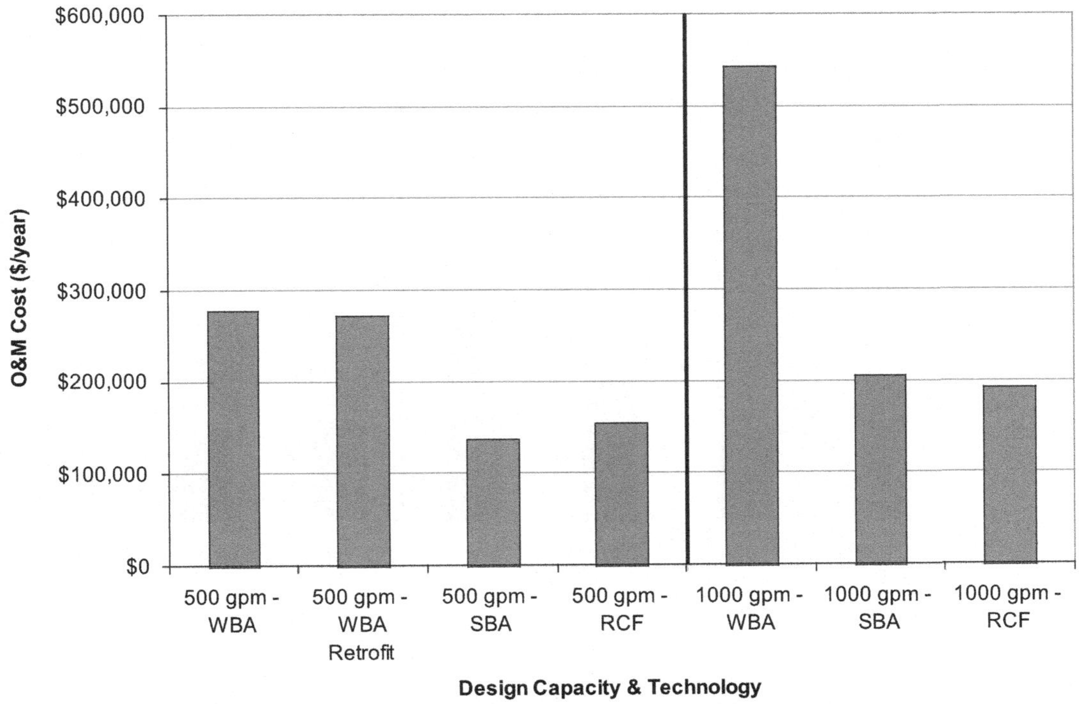

Figure 6.11 Comparison of annual O&M cost estimates for the three technologies at two flow rates

technology, the capital cost estimates were $2.9 million and $3.5 million for the 500 gpm and 1,000 gpm systems, respectively.

The figure highlights the relatively high capital costs associated with the RCF technology. RCF capital costs may be reduced by vendor-specific design and competition during a bidding process.

SBA and WBA applications had similar capital costs. However, the potential retrofit of existing GAC contactors at the Glendale GS-3 well site could significantly reduce the capital cost by approximately $700,000.

O&M Cost

Annual O&M cost estimates are presented in Figure 6.11. For the WBA technology, the annual O&M cost estimates were $270,000, $280,000, and $540,000 for the retrofitted 500 gpm, new 500 gpm, and 1,000 gpm systems, respectively. For the SBA technology, the annual O&M cost estimates were $140,000 and $210,000 for the 500 gpm and 1,000 gpm systems, respectively. For the RCF technology, the annual O&M cost estimates were $150,000 and $190,000 for the 500 gpm and 1,000 gpm systems, respectively.

In general, SBA and RCF had lower O&M costs than WBA resin. The high cost of resin replacement and acid for the WBA system eclipsed the residuals disposal costs that contributed much of the SBA and RCF O&M costs.

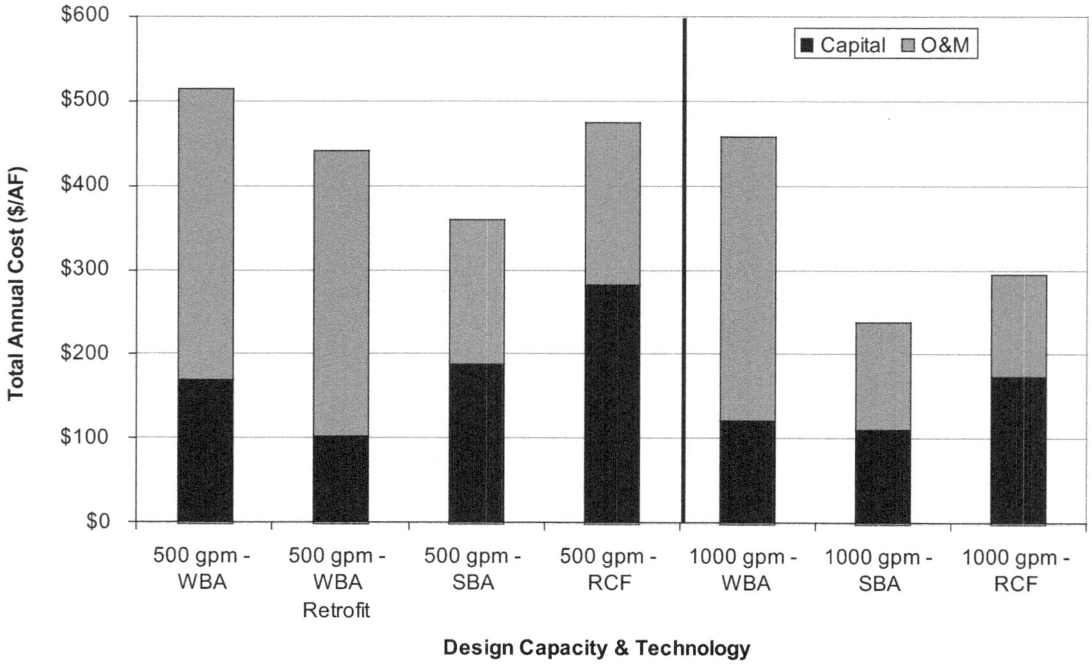

Figure 6.12 Comparison of annualized cost estimates for the three technologies at two flow rates

Annualized Cost

Figure 6.12 shows an annualized cost comparison of the various technologies. Detailed annualized cost estimates are also listed in Table 6.3.

For the 500 gpm system, the SBA option appears to be the lowest annualized cost at $360 per acre-foot (AF). However, concerns exist for the long-term disposal of brine generated by the SBA system. The retrofitted 500 gpm WBA system has the second lowest annualized cost ($440/AF), which is a viable choice if a 500 gpm system is to be installed.

For the 1,000 gpm system, the SBA process has the lowest annualized cost at $240, followed by RCF at $300/AF.

SUMMARY

Refined cost estimates were developed for three chromium treatment technologies (SBA, WBA, and RCF) using vendor-based estimates and standard cost estimation practices. For each technology, the following primary cost drivers were identified:

- Siemens/ Rohm & Haas WBA (Duolite A7 resin): Approximately 67% of the annualized costs were O&M costs, including resin replacement and acid for pH depression. This technology is particularly advantageous as a once-through treatment system with only solid residuals (i.e., no brine). Potential exists for the resin cost to be reduced, since costs were significantly lower 2 years prior to this cost update.
- BasinWater SBA: BasinWater system feasibility is predicated on several key unknowns. First, the ability to regenerate the resin many times has been called into

Table 6.3
Annualized treatment cost estimates for Cr(VI) removal

		Annualized costs ($/AF)		
Technology	Flow (gpm)	Capital	O&M	Total
WBA	500 (Retrofit)	100	340	440
	500	170	350	520
	1,000	120	340	460
SBA	500	190	170	360
	1,000	110	130	240
RCF	500	280	190	470
	1,000	180	120	300

question by Siemens' refusal to sell regenerable SBA technology for chromate treatment due to loss of resin capacity after multiple regenerations. Second, brine processing yielding a non-hazardous liquid waste and hazardous solid waste is critical to providing the low costs shown in this evaluation. Inability to make the brine non-hazardous would render the technology significantly more expensive, as shown in the previous Phase II cost evaluation (MEC 2005). Finally, the likelihood of future brine disposal to a sewer connection introduces a large unknown to the process.

- RCF: Annualized RCF costs were split between capital costs, operating costs, and residuals costs. Residuals disposal contributed the largest annual cost, followed by capital costs due to the need for filters and backwash handling and treatment. Other filtration strategies, such as microfiltration, may offer cost savings in the filtration capital costs. Further testing would be necessary to determine the effectiveness of alternate filtration strategies.

For the 500-gpm case, a retrofit of the GAC contactors at Glendale GS-3 well site for WBA is the most cost-effective option if SBA is judged to be unattractive due to brine disposal needs. Likewise, RCF would be the most cost-effective treatment solution for a 1,000 gpm installation if the risks associated with the SBA are not acceptable.

CHAPTER 7
EXPERT PANEL WORKSHOP

Following completion of pilot testing and cost updates, an expert panel workshop for the Phase III Bridge Project was held on October 12, 2006 at the City of Glendale Council Chambers. The expert panel meeting was co-hosted by the USEPA, Glendale Water and Power, and AwwaRF.

The expert panel members in attendance at the workshop included Dr. Pankaj Parekh from the Los Angeles Department of Water and Power (LADWP), Dr. Sun Liang from the Metropolitan Water District of Southern California, Dr. Bruce Macler from the USEPA, Dr. Richard Sakaji from California DHS, Dr. Irwin (Mel) Suffet from UCLA, Dr. Laurie McNeill from Utah State University, Dr. Arup SenGupta from Lehigh University, and Dr. Gary Amy from the United Nations Educational, Scientific, and Cultural Organization (UNESCO; attending via teleconference). The panel discussion was moderated by project manager Traci Case from AwwaRF.

The expert panel meeting was open to the public. More than 30 people interested in Cr(VI) issues attended the event. In addition, the City of Glendale broadcasted the meeting live on Glendale public television and over the internet as streaming media.

CHARGE TO THE EXPERT PANEL

Glendale's charge to the expert panel was to identify cost-effective Cr(VI) treatment technologies that are appropriate for further testing at demonstration scale (approximately 500 gpm or 1,000 gpm treatment capacity) based on the technical information presented at the meeting.

The panel was asked to consider the following criteria in the evaluation process:

- Technology maturity
- Probable success in Glendale and elsewhere
- Cost of the treatment facilities
- Ease of operations and maintenance, including future reliability of the treatment processes
- Required permitting and approval processes

CHROMIUM (VI) TREATMENT TECHNOLOGIES CONSIDERED

Research results from the Phase I Bench-Scale Study, Phase II Pilot-Scale Study, and Phase III Bridge Project were presented to the expert panel by Dr. Michael McGuire and Dr. Nicole Blute. Based on the studies' findings, three technologies emerged as leading candidates for consideration in demonstration testing:

- Reduction/Coagulation/Filtration using ferrous sulfate
- Fixed Bed Weak-Base Anion Exchange Resin with constant pH control
- Fixed Bed Strong-Base Anion Exchange Resin with brine treatment

SUMMARY OF THE EXPERT PANEL DISCUSSIONS

Expert panel discussions about the three technologies are summarized below according to the technology.

Reduction/Coagulation/Filtration

Brief Description

During the RCF process, Cr(VI) is first reduced to Cr(III) with the addition of excess ferrous iron (Fe^{2+}), which is oxidized to ferric iron (Fe^{3+}). Cr(III) then either precipitates or forms a co-precipitate with the ferric iron. The ferric iron/Cr(III) particles form a larger floc during the coagulation (aeration) stage. Particles are then removed by a dual-media filter (or other filter, such as a microfiltration membrane) in the final step.

Advantages

The expert panel generally favored this technology for the following reasons:

- The mechanism of RCF treatment is fully understood
- RCF is a proven technology for the application of Cr(VI) removal, as evidenced by the successful operation of a similar system at Topock, California
- RCF can be optimized during the demonstration-scale study to accommodate potential increases in Cr(VI) concentrations in the influent water
- California DHS permitting will likely be easier for large systems using RCF compared to the other two technologies

Disadvantages

The panel expressed the following disadvantages of the RCF system:

- The capital cost of constructing the RCF system is very high (preliminary cost estimates indicated that the construction of a 500 gpm RCF system could cost $2.8 million, and a 1,000 gpm system could cost $3.5 million).
- RCF may require frequent operator oversight and continuous monitoring to optimize the removal of Cr(VI).
- In California, a related treatment process (i.e., coagulation/filtration) is only an accepted best available technology for arsenic removal in systems with greater than 500 service connections due to operational complexities.

Weak-Base Anion Exchange

Brief Description

The mechanism of Cr(VI) removal by WBA resins was originally believed to be similar to that of SBA resins, except that the WBA resins are only useful in the acidic pH range where the functional groups are protonated and thus act as positively charged exchange sites to attract Cr(VI) (as chromate or bichromate ions). However, the WBA resin (Duolite A7 resin provided by Rohm & Haas) tested in the Phase II Pilot-Scale Study showed a much greater Cr(VI) removal capacity compared with all of the other SBA resins tested (i.e., approximately 20 times higher capacity). Other observations, such as leakage of Cr(III) during periods of low pH, indicated that another mechanism besides ion exchange may contribute to the high capacity of the WBA resins.

As part of the Phase III Bridge Project, several WBA resins were tested to assess capacity at equilibrium and under flow-through conditions. The impact of pH on capacity was also evaluated. Duolite A7 resin showed a high Cr(VI) capacity along with another WBA resin (ResinTech SIR-700, which did not perform quite as well as the Duolite A7 initially but improved over time). Laboratory studies using x-ray absorption spectroscopy confirmed that at least 95% of the chromium retained on both resins was present as Cr(III), whereas the influent water contained Cr(VI). So far, the true mechanism of Cr(VI) removal and retention by the WBA resins has not been fully understood but is known to involve a reduction process.

Advantages

The expert panel discussed the advantages of the WBA system, including the following:

- WBA resins have demonstrated a high Cr(VI) removal capacity (approximately 20 times higher than the conventional SBA resins tested)
- The operation of WBA system is comparatively easy, especially for a small system
- The WBA resins would be used in a single-pass, disposable mode, eliminating the need for resin regeneration with brine
- The WBA system can absorb fluctuations in influent Cr(VI) concentrations, although resin replacement would be more frequent at higher influent concentrations

Disadvantages or Uncertainties

The WBA resin was the most thoroughly discussed technology during the expert panel meeting, primarily because the mechanism for Cr(VI) removal is not fully understood. Pilot studies have indicated that besides ion exchange, redox reactions and complexation could also play a role in Cr(VI) removal by the WBA resins. One panel member raised the question: "Do we want to select a technology where the mechanism is not understood?"

The expert panel also expressed other concerns regarding the WBA system besides an incomplete understanding of the mechanism, including:

- Potential nitrosamine (including NDMA) leaching from resins
- Potential for formation/release of organic resin byproducts (e.g. formaldehyde or phenol, which are resin constituents in at least one of the WBA resins)
- Taste and odor issues related to the use of the resins if formaldehyde or phenol are released
- Cost of the WBA system (the highest annual O&M cost among the three technologies due to resin prices)
- The potential need to pre-condition the resins, which may explain the improved removals over time for the SIR-700 resin

- Establishing operational permit provisions is more difficult if the removal mechanism is not well understood. In turn, difficulties in establishing permit provisions may make the process of obtaining an operational permit more challenging
- The need for influent water pH reduction

Strong-Base Anion Exchange

Brief Description

Ion exchange with SBA resins is a commonly used technology in drinking water treatment for anion removal. Cr(VI) is retained on the SBA resin (as chromate or bichromate ion) by exchanging with chloride previously bound to the resin. SBA resins may be reused by regenerating the resins with concentrated brine (salt) solutions. Pilot testing demonstrated up to seven regeneration cycles using recycled brine.

BasinWater provided a proposal for a regenerable SBA system for Cr(VI) treatment. By contrast, Siemens did not provide a follow-up proposal since they no longer market SBA regeneration applications for Cr(VI) removal. During recent testing in Kansas, Siemens encountered difficulty in regenerating the resin to full capacity after approximately 12 to 15 regeneration cycles. Consequently, Siemens now offers only the WBA system and a single-pass SBA system.

Advantages

The advantages of the SBA system discussed by the expert panel included:

- SBA is an established technology for other contaminants and the mechanism is well understood.
- The overall Cr(VI) treatment cost using the SBA system (including capital and O&M cost) is the least among the three technologies.
- The SBA system can absorb fluctuations in influent Cr(VI) concentrations.

Disadvantages

Disadvantages and uncertainties associated with the SBA system at this time were discussed by the expert panel, including:

- Brine disposal: The high concentration of total dissolved solids and chloride in the brine may ultimately prevent its discharge into the sanitary sewer systems. At the GWTP site, the Los Angeles-Glendale Water Reclamation Plant will not accept high chloride brine. For the GS-3 well site (which is located in Los Angeles), BasinWater recommended truck disposal of non-hazardous brine into a sewer hookup leading to the Los Angeles Hyperion Wastewater Treatment Plant. However, the availability of this connection into the future is not guaranteed.
- The quantity of brine BasinWater claims to generate is orders of magnitude lower than other SBA technologies in the area for removal of other contaminants, which either reflects a high degree of efficiency with the BasinWater system or a lower-than-actual estimate of brine production.

- Uncertainty exists regarding the ability to repeatedly regenerate SBA resin to near its original capacity for Cr(VI) removal.

EXPERT PANEL RECOMMENDATIONS

At the end of the meeting, the expert panel provided the following overall recommendations:

- All panel members recommended that the RCF system be tested in the demonstration-scale study.
- All panel members recommended that the mechanism of the WBA resins be thoroughly investigated in additional bench-scale studies. The understanding of the WBA mechanism will not only help Glendale in the demonstration-scale study but will also provide a new solution to other water utilities needing Cr(VI) treatment.
- The SBA system should not be further tested at demonstration scale.

CHAPTER 8
SUMMARY AND CONCLUSIONS

SUMMARY

The City of Glendale has devoted significant resources to the search for effective Cr(VI) removal technologies for drinking water treatment to achieve low treatment goals. During the extensive Phase I Bench-Scale Study and Phase II Pilot-Scale Study, WBA resins demonstrated an unexpectedly high Cr(VI) removal capacity. Before the WBA resin was considered for further testing at demonstration scale, the Phase III Bridge Project was conducted to investigate the efficiency of WBA resins under constant pH conditions and to compare the economics of their one-time use with the other two candidates: SBA and RCF.

The Phase III Bridge Project first screened six promising WBA resins for Cr(VI) removal using isotherm tests. SIR-700 and Duolite A7 resin capacities for Cr(VI) exceeded those of the other four WBA resins and advanced to mini-column and pilot-column tests. Of note from the bench-scale evaluation was that both resins demonstrated slow kinetics to reach equilibrium, requiring more than 64 days. The high Cr(VI) capacities coupled with the slow kinetics of the SIR-700 and Duolite A7 resins indicated that a mechanism other than ion exchange could be involved in Cr(VI) removal.

The impact of pH on Cr(VI) removal by the SIR-700 and Duolite A7 resins was inconclusive in bench-scale testing and was thus further investigated in flow-through mini-columns. Mini-column testing results at pH values ranging from 5.6 to 7.2 showed different breakthrough characteristics for SIR-700 and Duolite A7 resins. However, both resins favored a lower pH for Cr(VI) removal, with a pH of 6.0 shown to be effective for both resins during the short mini-column testing period.

Another important finding from the mini-column evaluation was that the water from the Glendale GS-3 well was supersaturated with CO_2 and had a natural pH around 6.8. This ambient pH was considerably lower than the historical laboratory-reported pH data of 7.2–7.3 and translates into a lower acid addition cost if the WBA system is operated without breaking head from the well (i.e., the recommended treatment strategy).

SIR-700 and Duolite A7 resins were subsequently tested in pilot-scale columns at both the ambient pH and a pH of 6.0 to assess Cr(VI) capacity under typical operating conditions. Chromium breakthrough curves of both resins confirmed the importance of maintaining a pH of 6.0 for Cr(VI) removal. At pH 6.0, the Duolite A7 resin could treat approximately 45,000 BV of water before 5 µg/L total Cr and Cr(VI) concentrations were observed in the effluent.

Although the SIR-700 resin showed early breakthrough of concentrations greater than 5 µg/L at 2,200 BV and pH 6.0, Cr(VI) removal improved during the testing period. In fact, effluent total Cr and Cr(VI) concentrations were less than 5 µg/L at the end of the pilot-scale testing (after treating approximately 113,000 BV of water). The improvement in Cr(VI) removal by the SIR-700 resin was contrary to a typical ion exchange breakthrough curve, which may suggest that another removal mechanism was dominant or that the resin needed to be pre-conditioned.

Once the pilot testing was complete, resin residuals were tested for hazardous waste characteristics that would determine disposal options. Spent SIR-700 and Duolite A7 resins operated at pH 6.0 passed the federal TCLP test but failed the California WET evaluation based on chromium concentrations; these resins would thus be considered hazardous waste for disposal in California.

Low levels of uranium present in the groundwater was also found to accumulate on both spent resins over the extended period of resin operation. The spent Duolite A7 resins exceeded a total uranium concentration of 500 µg/g (the trigger for low-level radioactive waste designation) after treating approximately 113,000 BV of water. For Glendale, the operational life of the Duolite A7 resin would have to be limited to avoid generating a low-level radioactive waste.

Resin residuals were also studied extensively in the Bridge Project to explore the mechanisms for Cr(VI) removal. XRF analysis revealed that significant amounts of chromium, sulfur, vanadium, and copper were accumulated on both spent resins. No crystalline precipitates were found on the resins using XRD, indicating that the chromium on the resin is predominantly complexed or present as an amorphous phase. The most significant finding from the residuals analyses using XANES was that the primary chromium species retained on both resin residuals was Cr(III). The XANES analysis provided direct evidence that the reduction of Cr(VI) to Cr(III) was an important part of the Cr(VI) removal mechanism by the SIR-700 and Duolite A7 resins.

In addition to the technical evaluations of the WBA resins, cost estimates of demonstration-scale application were developed and compared with the other two promising Cr(VI) removal technologies: SBA and RCF. For the City of Glendale, the annualized Cr(VI) treatment costs were estimated at $440/AF for a 500 gpm WBA system using retrofitted GAC contactors at the GS-3 well site. The annualized cost estimates were $520/AF and $460/AF for new 500 gpm and 1,000 gpm WBA systems, respectively.

Although the annualized cost estimates were the lowest for the SBA option at both flow rates (500 gpm and 1,000 gpm), the uncertainty associated with future brine disposal could make the SBA process cost-prohibitive. At 500 gpm, the retrofitted WBA system was considered as a cost-effective Cr(VI) treatment technology for Glendale. For a 1,000 gpm system, the RCF process had the advantage of a lower annualized cost estimate of $300/AF compared to $460/AF of the 1,000 gpm WBA system.

CONCLUSIONS

Major conclusions from the Phase III Bridge Project include:

- Two WBA resins (ResinTech SIR-700 and Siemens/R&H Duolite A7) demonstrated high Cr(VI) removal capacities.
- At a pH of 6.0, total Cr and Cr(VI) concentrations in the Duolite A7 resin effluent remained lower than 5 µg/L up to 45,000 BV of Glendale GS-3 well water treated (containing 35 – 40 µg/L Cr(VI)), which confirmed the previous Phase II Pilot-Scale Study finding that the Duolite A7 resin was effective for Cr(VI) removal at acidic pH.
- Duolite A7 WBA resin offers an effective option for Cr(VI) removal in Glendale groundwater at pH 6.0.
- Spent SIR-700 and Duolite A7 resins would be considered as hazardous waste in California.
- The operational life of the Duolite A7 resin in Glendale groundwater would have to be limited to avoid excessive uranium accumulation (i.e., greater than 500 µg/g) and the triggering of low-level mixed radioactive waste regulations.
- The mechanisms of Cr(VI) removal by the SIR-700 and Duolite A7 resins are still not fully understood but the reduction of Cr(VI) to Cr(III) has been shown to be an important part of chromium retention by the resins.

- A retrofitted WBA system (using the Duolite A7 resin) was considered cost-effective for Cr(VI) removal in Glendale groundwater at 500 gpm, while the RCF system was deemed more cost-effective at 1,000 gpm.
- After reviewing the technical and cost information from the Phase I Bench-Scale Study, Phase II Pilot-Scale Study, and the Phase III Bridge Project, an expert panel concluded that the RCF system should be tested in the Phase III Demonstration-Scale Study.
- The expert panel also recommended that the Cr(VI) removal mechanisms by the WBA resins be further investigated in select bench-scale studies.
- No further testing of the SBA resin was advised by the expert panel.

NEXT STEP

Although the expert panel recommended RCF testing in the Phase III Demonstration-Scale Study, the high capital costs associated with this technology compared to WBA resins caused the City of Glendale to select a two-pronged approach to achieve the most reduction in the water supply hexavalent chromium levels for the lowest capital cost. A relatively inexpensive demonstration-scale WBA system (500 gpm) was selected for the next phase and will be accomplished by converting the existing two GAC contactors to ion exchange vessels at the Glendale GS-3 well site. Depending on the availability of sufficient funds, a smaller RCF system (e.g. 100 – 200 gpm) will be installed and tested for Cr(VI) removal from the GN wells at a location adjacent to the GWTP. In the meantime, additional bench-scale studies will be conducted to further elucidate the Cr(VI) removal mechanism by the WBA resins.

APPENDIX A
CONVERSION TABLE

SI equivalent units

	To convert		
Water distribution parameters	From customary units	To SI units	Multiply by
Concentrations	ppm	mg/L	—
	ppm	weight percent	1×10^{-4}
	ppb	µg/L	—
Hydraulic loading rate	gpm/ft^2	L/m^3/s	6.79×10^{-1}
Length	in	cm	2.54
	ft	m	3.048×10^{-1}
Pump capacity	gpm	m^3/s	6.309×10^{-5}
		or L/s	6.309×10^{-2}
Volume	gallons	m^3	3.785×10^{-3}
		or L	3.785
	ft^3	m^3	2.832×10^{-2}

APPENDIX B
TECHNOLOGY COST ESTIMATES, INCLUDING CAPITAL COST SUMMARIES, CAPITAL COST DETAILS, AND O&M COSTS

- Siemens/R&H Weak-Base Anion Exchange (at pH 6.0) – 500 gpm
- Siemens/R&H Weak-Base Anion Exchange (at pH 6.0) – 500 gpm Retrofit
- Siemens/R&H Weak-Base Anion Exchange (pH 6.0) – 1,000 gpm
- BasinWater Strong-Base Anion Exchange – 500 gpm
- BasinWater Strong-Base Anion Exchange – 1,000 gpm
- Reduction/Coagulation/Filtration – 500 gpm
- Reduction/Coagulation/Filtration – 1,000 gpm

Siemens/R&H Weak-Base Anion Exchange (pH 6.0) – 500 gpm:

City of Glendale
Capital Cost Estimate - Weak Base Anion (WBA) 500 gpm

Company: Malcolm Pirnie, Inc.
Project: Phase III Hexavalent Chromium Demonstration System
Submittal: Conceptual Design Level Cost Estimate
Work Task:

Date: 29-Sep-06
Estimator: SMD
Checker: GB, MJM, NKB
Cost Index: ENR CCI = 8572.47 (Los Angeles, September 2006)

Division Summary		Total
Division 1 - General Conditions		$50,000.00
Division 2 - Site Construction		$81,690.00
Division 3 - Concrete		$55,500.00
Division 4 - Masonry		$0.00
Division 5 - Metals		$15,500.00
Division 6 - Wood & Plastics		$0.00
Division 7 - Thermal & Moisture Protection		$250.00
Division 8 - Doors & Windows		$0.00
Division 9 - Finishes		$14,750.00
Division 10 - Specialties		$500.00
Division 11 - Equipment		$604,059.60
Division 12 - Furnishings		$0.00
Division 13 - Special Construction		$21,600.00
Division 14 - Conveying Systems		$0.00
Division 15 - Mechanical		$34,640.00
Division 16 - Electrical		$150,000.00
Division 17 - Instrumentation and Control		$115,000.00
	Division 1 - 17 Subtotal	**$1,143,489.60**
	Insurance @ 2.5%	**$28,587.24**
	Bonds @ 2.0%	**$22,869.79**
	Overhead & Profit @ 10%	**$114,348.96**
	Engineering @ 10%	**$114,348.96**
	Total	**$1,423,644.55**
	Contingency @ 20.0%	**$284,728.91**
	CONCEPTUAL LEVEL PROBABLE CONSTRUCTION COST	**$1,708,373.46**
	(September 2006)	

City of Glendale
Capital Cost Estimate - Weak Base Anion (WBA) 500 gpm

Company: Malcolm Pirnie, Inc.
Project: Phase III Hexavalent Chromium Demonstration System
Submittal: Conceptual Design level Cost Estimate
Work Task:

Date: 29-Sep-06
Estimator: SMD
Checker: GB, MJM, NKB
Cost Index: ENR CCI = 8572.47 (Los Angeles, September 2006)

	Specification Section	Unit	Quantity	Cost	Installation Factor	Total
Division 1 - General Conditions						
Div 1	General Conditions	LS	1	$ 50,000.00	1.0	$ 50,000.00
	Mobilization/Demobilization					
	Division 1 Total					**$50,000.00**
Division 2 - Site Construction						
02220	Site Preparation	LS	1	$ 5,000.00	1.0	$5,000.00
02230	Clearing					
02315	Excavation and Backfill					
	Excavation and grading of of demonstration facility limits	LS	1	$ 50,000.00	1.0	$50,000.00
	Excess Material to be removed	CY	1,570	$ 8.00	1.0	$12,560.00
	Backfill	CY	785	$ 18.00	1.0	$14,130.00
	Division 2 Total					**$81,690.00**
Division 3 - Concrete						
03100	Concrete Formwork			In 03300		
03300	Concrete					
	Equipment Slab (3,000-sf)	CY	111	$ 500.00	1.0	$55,500.00
	Division 3 Total					**$55,500.00**
Division 4 - Masonry						
Division 5 - Metals						
05051	Anchor Bolts, Toggle Bolts, and Concrete Inserts	LS	1	$ 500.00	1.0	$500.00
05501	Miscellaneous Metal Fabrications (includes access platforms)	LS	1	$ 15,000.00	1.0	$15,000.00
	Division 5 Total					**$15,500.00**
Division 6 - Wood & Plastics						
Division 7 - Thermal & Moisture Protection						
07920	Caulking and Sealants	LS	1	$ 250.00	1.0	$250.00
	Division 7 Total					**$250.00**
Division 8 - Doors & Windows						
Division 9 - Finishes						
09611	Concrete Hardener	SF	3,000	$ 3.25	1.0	$9,750.00
09900	Painting	LS	1	$ 5,000.00	1.0	$5,000.00
	Division 9 Total					**$14,750.00**

Specification Section		Unit	Quantity	Cost	Installation Factor	Total
Division 10 - Specialties						
10400	Identification Devices	LS	1	$ 500.00	1.0	$500.00
	Division 10 Total					**$500.00**
Division 11 - Equipment						
11179	Fiberglass Reinforced Plastic Tanks (HCl bulk Storage Tank)	LS	1	$ 5,000.00	1.2	$6,000.00
11180	Fiberglass Reinforced Plastic Tanks (Waste Equilization Tank)	LS	1	$ 12,500.00	1.2	$15,000.00
11216	Hydrochloric Acid Feed System w/ scrubber	LS	1	$ 60,482.00	1.2	$72,578.40
11193	Bag Filters	EA	2	$ 17,500.00	1.2	$42,000.00
11195	Fixed Bed Ion Exchange System (Weak Base Anion)	LS	1	$ 161,903.00	1.2	$194,283.60
11196	Weak Base Anion Resin	LS	1	$ 198,898.00	1.2	$238,677.60
11530	Pumps, General	EA	2	$ 14,800.00	1.2	$35,520.00
	Division 11 Total					**$604,059.60**
Division 12 - Furnishings						
Division 13 - Special Construction						
13125	FRP Walk-In Enclosure (MCC)	LS	1	$ 18,000.00	1.2	$21,600.00
	Division 13 Total					**$21,600.00**
Division 14 - Conveying Systems						
Division 15 - Mechanical						
15051	Buried Piping Installation					
	8" Influent Piping	FT	250	$ 21.00	1.4	$7,350.00
	Effluent Piping	FT	250	$ 21.00	1.4	$7,350.00
	Backwash Waste Piping	FT	100	$ 21.00	1.4	$2,940.00
15052	Exposed Piping Installation					
	Ion Exchange System Process Piping			In 11195		
15055	Pipe Hangers and Supports	LS	1	$ 5,000.00	1.0	$5,000.00
15061	Ductile Iron Pipe			In 15051		
15067	Thermoplastic Pipe			In 15051 & 15052		
15100	Valves, 4-Inch and Larger	LS	1	$ 10,000.00	1.2	$12,000.00
	Division 15 Total					**$34,640.00**
Division 16 - Electrical						
16050	General Provisions	LS	1	$ 150,000.00	1.0	$150,000.00
	Division 16 Total					**$150,000.00**
Division 17 - Instrumentation and Control						
17400	Instrumentation and Control	LS	1	$ 115,000.00	1.0	$ 115,000.00
	Division 17 Total					**$115,000.00**
					Division 1 - 17 Subtotal	**$1,143,489.60**

City of Glendale
Operation and Maintenance Costs - Weak Base Anion (WBA) 500 gpm

Company:	Malcolm Pirnie, Inc.	**Date:**	29-Sep-06
Project:	Phase III Hexavalent Chromium Demonstration System	**Estimator:**	SMD
Submittal:	Conceptual Design Level Cost Estimate	**Checker:**	GB, MJM, NKB
Work Task:		**Cost Index:**	ENR CCI = 8572.47 (Los Angeles, September 2006)

Item	Description	O&M Misc.	Energy	Chemicals	Waste Disposal	Total O&M	Notes
1.0	**Weak Base Anion System**						
1.1	Resin/Media Replacement	$175,404				$175,404.00	Assumed 100,000 bed-volume capacity
1.2	Pre-Filters filter replacement	$1,440				$1,440.00	Assumed monthly replacement of bag filters
1.3	Effluent Booster Pumps		$6,345			$6,344.53	Boosting IX effluent to aerator influent. 500-gpm @ 50' TDH
1.4	Chemicals			$69,259		$69,258.75	Feed water conditioning with 31% HCl. Feed Rate: 165 gpd of 31% HCl. Cost: $1.15/gal for 2,000-gal loads
1.5	Miscellaneous Maintenance	$12,081				$12,081.19	Assumed 2% of equipment costs (Div 11)
	Subtotal - WBA System					**$264,528.48**	
2.0	**Waste Disposal**						
2.1	Liquid Waste Disposal (Backwash Water)				$1,035	$1,035.00	Backwash waste to sanitary sewer based on 321,000 gallons liquid waste per year (for 500-gpm system)
2.2	Resin/Media Disposal					$0.00	Resin disposal costs included in resin replacement cost (if considered separately, cost would be $1675 for 6.7 tons at $445/ton)
	Subtotal - Waste Disposal					**$1,035.00**	
3.0	**Labor**						
3.1	Labor	$12,500				$12,500.00	Labor based 0.125 FTE at $100,000/yr.
	Subtotal - Labor					**$12,500.00**	
TOTAL	**Subtotals**	$201,425	$6,345	$69,259	$1,035		
	Total					**$278,063.48**	

Siemens/R&H Weak-Base Anion Exchange (pH 6.0) – 500 gpm Retrofit:

City of Glendale
Capital Cost Estimate - Weak Base Anion (WBA) 500 gpm Retrofit

Company: Malcolm Pirnie, Inc.
Project: Phase III Hexavalent Chromium Demonstration System
Submittal: Conceptual Design Level Cost Estimate
Work Task:

Date: 29-Sep-06
Estimator: SMD
Checker: GB, MJM, NKB
Cost Index: ENR CCI = 8572.47 (Los Angeles, September 2006)

Division Summary		Total
Division 1 - General Conditions		$25,000.00
Division 2 - Site Construction		$5,000.00
Division 3 - Concrete		$9,500.00
Division 4 - Masonry		$0.00
Division 5 - Metals		$15,500.00
Division 6 - Wood & Plastics		$0.00
Division 7 - Thermal & Moisture Protection		$0.00
Division 8 - Doors & Windows		$0.00
Division 9 - Finishes		$6,625.00
Division 10 - Specialties		$0.00
Division 11 - Equipment		$322,759.40
Division 12 - Furnishings		$0.00
Division 13 - Special Construction		$21,600.00
Division 14 - Conveying Systems		$0.00
Division 15 - Mechanical		$25,820.00
Division 16 - Electrical		$150,000.00
Division 17 - Instrumentation and Control		$115,000.00
	Division 1 - 17 Subtotal	**$696,804.40**
	Insurance @ 2.5%	**$17,420.11**
	Bonds @ 2.0%	**$13,936.09**
	Overhead & Profit @ 10%	**$69,680.44**
	Engineering @ 10%	**$69,680.44**
	Total	**$867,521.48**
	Contingency @ 20.0%	**$173,504.30**
	CONCEPTUAL LEVEL PROBABLE CONSTRUCTION COST	**$1,041,025.77**
	(September 2006)	

City of Glendale

Capital Cost Estimate - Weak Base Anion (WBA) Retrofit 500 gpm Retrofit

Company: Malcolm Pirnie, Inc.
Project: Phase III Hexavalent Chromium Demonstration System
Submittal: Conceptual Design Level Cost Estimate
Work Task:

Date: 29-Sep-06
Estimator: SMD
Checker: GB, MJM, NKB
Cost Index: ENR CCI = 8572.47 (Los Angeles, September 2006)

Specification Section		Unit	Quantity	Cost	Installation Factor	Total
Division 1 - General Conditions						
Div 1	General Conditions	LS	1	$ 25,000.00	1.0	$ 25,000.00
	Mobilization/Demobilization					
	Division 1 Total					**$25,000.00**
Division 2 - Site Construction						
02220	Site Preparation	LS	1	$ 5,000.00	1.0	$5,000.00
	Division 2 Total					**$5,000.00**
Division 3 - Concrete						
03100	Concrete Formwork			In 03300		
03300	Concrete					
	Equipment Slab (500-sf)	CY	19	$ 500.00	1.0	$9,500.00
	Division 3 Total					**$9,500.00**
Division 4 - Masonry						
Division 5 - Metals						
05051	Anchor Bolts, Toggle Bolts, and Concrete Inserts	LS	1	$ 500.00	1.0	$500.00
05501	Miscellaneous Metal Fabrications (includes access platforms)	LS	1	$ 15,000.00	1.0	$15,000.00
	Division 5 Total					**$15,500.00**
Division 6 - Wood & Plastics						
Division 7 - Thermal & Moisture Protection						
Division 8 - Doors & Windows						
Division 9 - Finishes						
09611	Concrete Hardener	SF	500	$ 3.25	1.0	$1,625.00
09900	Painting	LS	1	$ 5,000.00	1.0	$5,000.00
	Division 9 Total					**$6,625.00**

Specification Section		Unit	Quantity	Cost	Installation Factor	Total
Division 10 - Specialties						
Division 11 - Equipment						
11179	Fiberglass Reinforced Plastic Tanks (HCl bulk Storage Tank)	LS	1	$ 5,000.00	1.2	$6,000.00
	Fiberglass Reinforced Plastic Tanks (Waste Equilization Tank)	LS	1	$ 12,500.00	1.2	$15,000.00
11216	Hydrochloric Acid Feed System w/ scrubber	LS	1	$ 60,482.00	1.2	$72,578.40
11195	Fixed Bed Ion Exchange System (Weak Base Anion), includes prefilter housings, filters, vessel conversion labor, media loading labor and start-up assistance.	LS	1	$ 193,661.00	1.0	$193,661.00
11196	Weak Base Anion Resin	LS	1	$ -	1.0	$0.00
11530	Pumps, General (Effluent Booster Pumps)	EA	2	$ 14,800.00	1.2	$35,520.00
	Division 11 Total					**$322,759.40**
Division 12 - Furnishings						
Division 13 - Special Construction						
13125	FRP Walk-In Enclosure (Electrical gear/MCC)	LS	1	$ 18,000.00	1.2	$21,600.00
	Division 13 Total					**$21,600.00**
Division 14 - Conveying Systems						
Division 15 - Mechanical						
15051	Buried Piping Installation					
	8" Influent Piping	FT	100	$ 21.00	1.4	$2,940.00
	Effluent Piping	FT	100	$ 21.00	1.4	$2,940.00
	Backwash Waste Piping	FT	100	$ 21.00	1.4	$2,940.00
15052	Exposed Piping Installation					
	Ion Exchange System Process Piping			In 11195		
15055	Pipe Hangers and Supports	LS	1	$ 5,000.00	1.0	$5,000.00
15061	Ductile Iron Pipe			In 15051		
15067	Thermoplastic Pipe			In 15051 & 15052		
15100	Valves, 4-Inch and Larger	LS	1	$ 10,000.00	1.2	$12,000.00
	Division 15 Total					**$25,820.00**
Division 16 - Electrical						
16050	General Provisions	LS	1	$ 150,000.00	1.0	$150,000.00
	Division 16 Total					**$150,000.00**
Division 17 - Instrumentation and Control						
17400	Instrumentation and Control	LS	1	$ 115,000.00	1.0	$ 115,000.00
	Division 17 Total					**$115,000.00**
					Division 1 - 17 Subtotal	**$696,804.40**

City of Glendale
Operation and Maintenance Costs - Weak Base Anion (WBA) Retrofit

Company:	Malcolm Pirnie, Inc.	**Date:**	29-Sep-06
Project:	Phase III Hexavalent Chromium Demonstration System	**Estimator:**	SMD
Submittal:	Conceptual Design Level Cost Estimate	**Checker:**	GB, MJM, NKB
Work Task:		**Cost Index:**	ENR CCI = 8572.47 (Los Angeles, September 2006)

Item	Description	O&M Misc.	Energy	Chemicals	Waste Disposal	Total O&M	Notes
1.0	**Weak Base Anion System**						
1.1	Resin/Media Replacement	$175,404				$175,404.00	Assumed 100,000 bed-volume capacity
1.2	Pre-Filters filter replacement	$1,440				$1,440.00	Assumed monthly replacement of bag filters
1.3	Effluent Booster Pumps		$6,345			$6,344.53	Boosting IX effluent to aerator influent. 500-gpm @ 50' TDH
1.4	Chemicals			$69,259		$69,258.75	Feed water conditioning with 31% HCl. Feed Rate: 165 gpd of 31% HCl. Cost: $1.15/gal for 2,000-gal loads
1.5	Miscellaneous Maintenance	$6,455				$6,455.19	Assumed 2% of equipment costs (Div 11)
	Subtotal - WBA System					**$258,902.47**	
2.0	**Waste Disposal**						
2.1	Liquid Waste Disposal (Backwash Water)				$1,035	$1,035.00	Backwash waste based on 321,000 gallons liquid waste (for 500-gpm system)
2.2	Resin/Media Disposal					$0.00	Resin disposal costs included in resin replacement cost (if considered separately, cost would be $1675 for 6.7 tons at $445/ton)
	Subtotal - Waste Disposal					**$1,035.00**	
3.0	**Labor**						
3.1	Labor	$12,500				$12,500.00	Labor based 0.125 FTE at $100,000/yr.
	Subtotal - Labor					**$12,500.00**	
TOTAL	**Subtotals**	$195,799	$6,345	$69,259	$1,035		
	Total					**$272,437.47**	

Siemens/R&H Weak-Base Anion Exchange (pH 6.0) – 1,000 gpm:

City of Glendale
Capital Cost Estimate - Weak Base Anion (WBA) 1000 gpm

Company:	Malcolm Pirnie, Inc.	**Date:**	29-Sep-06
Project:	Phase III Hexavalent Chromium Demonstration System	**Estimator:**	SMD
Submittal:	Conceptual Design Level Cost Estimate	**Checker:**	GB, MJM, NKB
Work Task:		**Cost Index:**	ENR CCI = 8572.47 (Los Angeles, September 2006)

Division Summary		Total
Division 1 - General Conditions		$75,000.00
Division 2 - Site Construction		$81,690.00
Division 3 - Concrete		$62,500.00
Division 4 - Masonry		$0.00
Division 5 - Metals		$15,500.00
Division 6 - Wood & Plastics		$0.00
Division 7 - Thermal & Moisture Protection		$0.00
Division 8 - Doors & Windows		$0.00
Division 9 - Finishes		$16,375.00
Division 10 - Specialties		$500.00
Division 11 - Equipment		$947,127.60
Division 12 - Furnishings		$0.00
Division 13 - Special Construction		$21,600.00
Division 14 - Conveying Systems		$0.00
Division 15 - Mechanical		$50,000.00
Division 16 - Electrical		$210,000.00
Division 17 - Instrumentation and Control		$165,000.00
	Division 1 - 17 Subtotal	**$1,645,292.60**
	Insurance @ 2.5%	**$41,132.32**
	Bonds @ 2.0%	**$32,905.85**
	Overhead & Profit @ 10%	**$164,529.26**
	Engineering @ 10%	**$164,529.26**
	Total	**$2,048,389.29**
	Contingency @ 20.0%	**$409,677.86**
	CONCEPTUAL LEVEL PROBABLE CONSTRUCTION COST	**$2,458,067.14**
	(September 2006)	

City of Glendale
Capital Cost Estimate - Weak Base Anion (WBA) 1000 gpm

Company: Malcolm Pirnie, Inc.
Project: Phase III Hexavalent Chromium Demonstration System
Submittal: Conceptual Design level Cost Estimate
Work Task:

Date: 29-Sep-06
Estimator: SMD
Checker: GB, MJM, NKB
Cost Index: ENR CCI = 8572.47 (Los Angeles, September 2006)

Specification Section		Unit	Quantity	Cost	Installation Factor	Total
Division 1 - General Conditions						
Div 1	General Conditions	LS	1	$ 75,000.00	1.0	$75,000.00
	Mobilization/Demobilization					
	Division 1 Total					**$75,000.00**
Division 2 - Site Construction						
02220	Site Preparation	LS	1	$ 5,000.00	1.0	$5,000.00
02230	Clearing					
02315	Excavation and Backfill					
	Excavation and grading of of demonstration facility limits	LS	1	$ 50,000.00	1.0	$50,000.00
	Excess Material to be removed	CY	1,570	$ 8.00	1.0	$12,560.00
	Backfill	CY	785	$ 18.00	1.0	$14,130.00
	Division 2 Total					**$81,690.00**
Division 3 - Concrete						
03100	Concrete Formwork			In 03300		
03300	Concrete					
	Equipment Slab (3,500-sf)	CY	125	$ 500.00	1.0	$62,500.00
	Division 3 Total					**$62,500.00**
Division 4 - Masonry						
Division 5 - Metals						
05051	Anchor Bolts, Toggle Bolts, and Concrete Inserts	LS	1	$ 500.00	1.0	$500.00
05501	Miscellaneous Metal Fabrications (includes access platform)	LS	1	$ 15,000.00	1.0	$15,000.00
	Division 5 Total					**$15,500.00**
Division 6 - Wood & Plastics						
Division 7 - Thermal & Moisture Protection						
Division 8 - Doors & Windows						
Division 9 - Finishes						
09611	Concrete Hardener	SF	3,500	$ 3.25	1.0	$11,375.00
09900	Painting	LS	1	$ 5,000.00	1.0	$5,000.00
	Division 9 Total					**$16,375.00**

Specification Section		Unit	Quantity	Cost	Installation Factor	Total
Division 10 - Specialties						
10400	Identification Devices	LS	1	$ 500.00	1.0	$500.00
	Division 10 Total					**$500.00**
Division 11 - Equipment						
11179	Fiberglass Reinforced Plastic Tanks (HCl bulk Storage Tank)	LS	1	$ 5,000.00	1.2	$6,000.00
	Fiberglass Reinforced Plastic Tanks (Waste Equilization Tank)	LS	1	$ 12,500.00	1.2	$15,000.00
11216	Hydrochloric Acid Feed System w/scrubber	LS	1	$ 80,184.00	1.2	$96,220.80
11193	Bag Filters	EA	2	$ 19,750.00	1.2	$47,400.00
11195	Fixed Bed Ion Exchange System (Weak Base Anion)	LS	1	$ 217,692.00	1.2	$261,230.40
11196	Weak Base Anion Resin	LS	1	$ 397,797.00	1.2	$477,356.40
15130	Pumps, General	EA	2	$ 18,300.00	1.2	$43,920.00
	Division 11 Total					**$947,127.60**
Division 12 - Furnishings						
Division 13 - Special Construction						
13125	FRP Walk-In Enclosure (MCC)	LS	1	$ 18,000.00	1.2	$21,600.00
	Division 13 Total					**$21,600.00**
Division 14 - Conveying Systems						
Division 15 - Mechanical						
15051	Buried Piping Installation					
	12" Influent Piping	FT	250	$ 25.00	1.4	$8,750.00
	Effluent Piping	FT	250	$ 25.00	1.4	$8,750.00
	Backwash Waste Piping	FT	100	$ 25.00	1.4	$3,500.00
15052	Exposed Piping Installation					
	Ion Exchange System Process Piping			In 11195		
15055	Pipe Hangers and Supports	LS	1	$ 5,000.00	1.0	$5,000.00
15061	Ductile Iron Pipe			In 15051		
15067	Thermoplastic Pipe			In 15051 & 15052		
15100	Valves & Fittings (Misc.), 4-Inch and Larger	LS	0	$ 20,000.00	1.2	$24,000.00
	Division 15 Total					**$50,000.00**
Division 16 - Electrical						
16050	General Provisions	LS	1	$ 210,000.00	1.0	$210,000.00
	Division 16 Total					**$210,000.00**
Division 17 - Instrumentation and Control						
17400	Instrumentation and Control	LS	1	$ 165,000.00	1.0	$165,000.00
	Division 17 Total					**$165,000.00**
					Division 1 - 17 Subtotal	**$1,645,292.60**

City of Glendale

Operation and Maintenance Costs - Weak Base Anion (WBA) 1000 gpm

Company:	Malcolm Pirnie, Inc.	**Date:**	29-Sep-06
Project:	Phase III Hexavalent Chromium Demonstration System	**Estimator:**	SMD
Submittal:	Conceptual Design Level Cost Estimate	**Checker:**	GB, MJM, NKB
Work Task:		**Cost Index:**	ENR CCI = 8572.47 (Los Angeles, September 2006)

Item	Description	O&M Misc.	Energy	Chemicals	Waste Disposal	Total O&M	Notes
1.0	**Weak Base Anion System**						
1.1	Resin/Media Replacement	$350,808				$350,808.00	Assumed 100,000 bed-volume capacity
1.2	Pre-Filters filter replacement	$1,440				$1,440.00	Assumed monthly replacement of bag filters
1.3	Effluent Booster Pumps		$6,345			$6,344.53	Boosting IX effluent to aerator influent. 1,000-gpm @ 50' TDH
1.4	Chemicals			$138,518		$138,517.50	Feed water conditioning with 31% HCl. Feed Rate: 165 gpd of 31% HCl. Cost: $1.15/gal for 2,000-gal loads
1.5	Miscellaneous Maintenance	$18,943				$18,942.55	Assumed 2% of equipment costs (Div 11)
	Subtotal - WBA System					**$516,052.59**	
2.0	**Waste Disposal**						
2.1	Liquid Waste Disposal (Backwash Water)				$2,363	$2,363.00	Backwash waste based on 733,000 gallons liquid waste (for 1,000-gpm system)
2.2	Resin/Media Disposal					$0.00	Resin disposal costs included in replacement cost (if considered separately, cost would be $3350 for 13.4 tons at $445/ton)
	Subtotal - Waste Disposal					**$2,363.00**	
3.0	**Labor**						
3.1	Labor	$25,000				$25,000.00	Labor based on 0.25 FTE at $100,000/yr.
	Subtotal - Labor					**$25,000.00**	
TOTAL	**Subtotals**	$396,191	$6,345	$138,518	$2,363		
	Total					**$543,415.59**	

BasinWater Strong-Base Anion Exchange – 500 gpm:

City of Glendale
Capital Cost Estimate - Strong Base Anion (SBA) - 500 gpm

Company:	Malcolm Pirnie, Inc.	**Date:**	29-Sep-06
Project:	Phase III Hexavalent Chromium Demonstration System	**Estimator:**	SMD
Submittal:	Conceptual Design Level Cost Estimate	**Checker:**	GB, MJM, NB
Work Task:		**Cost Index:**	ENR CCI = 8572.47 (Los Angeles, September 2006)

Division Summary		Total
Division 1 - General Conditions		$50,000.00
Division 2 - Site Construction		$81,690.00
Division 3 - Concrete		$66,500.00
Division 4 - Masonry		$0.00
Division 5 - Metals		$15,500.00
Division 6 - Wood & Plastics		$0.00
Division 7 - Thermal & Moisture Protection		$250.00
Division 8 - Doors & Windows		$0.00
Division 9 - Finishes		$16,700.00
Division 10 - Specialties		$0.00
Division 11 - Equipment		$719,344.00
Division 12 - Furnishings		$0.00
Division 13 - Special Construction		$21,600.00
Division 14 - Conveying Systems		$0.00
Division 15 - Mechanical		$34,640.00
Division 16 - Electrical		$150,000.00
Division 17 - Instrumentation and Control		$120,000.00
	Division 1 - 17 Subtotal	**$1,276,224.00**
	Insurance @ 2.5%	**$31,905.60**
	Bonds @ 2.0%	**$25,524.48**
	Overhead & Profit @ 10%	**$127,622.40**
	Engineering @ 10%	**$127,622.40**
	Total	**$1,588,898.88**
	Contingency @ 20.0%	**$317,779.78**
	CONCEPTUAL LEVEL PROBABLE CONSTRUCTION COST	**$1,906,678.66**
	(September 2006)	

City of Glendale
Capital Cost Estimate - Strong Base Anion (SBA) - 500 gpm

Company: Malcolm Pirnie, Inc.
Project: Phase III Hexavalent Chromium Demonstration System
Submittal: Conceptual Design Level Cost Estimate
Work Task:

Date: 29-Sep-06
Estimator: SMD
Checker: GB, MJM, NB
Cost Index: ENR CCI = 8572.47 (Los Angeles, September 2006)

Specification Section		Unit	Quantity	Cost	Installation Factor	Total
Division 1 - General Conditions						
Div 1	General Conditions	LS	1	$ 50,000.00	1.0	$ 50,000.00
	Mobilization/Demobilization					
	Division 1 Total					**$50,000.00**
Division 2 - Site Construction						
02220	Site Preparation	LS	1	$ 5,000.00	1.0	$5,000.00
02230	Clearing					
02315	Excavation and Backfill					
	Excavation and grading of of demonstration facility limits	LS	1	$ 50,000.00	1.0	$50,000.00
	Excess Material to be removed	CY	1,570	$ 8.00	1.0	$12,560.00
	Backfill	CY	785	$ 18.00	1.0	$14,130.00
	Division 2 Total					**$81,690.00**
Division 3 - Concrete						
03100	Concrete Formwork			In 03300		
03300	Concrete					
	Equipment Slab	CY	133	$ 500.00	1.0	$66,500.00
	Division 3 Total					**$66,500.00**
Division 4 - Masonry						
Division 5 - Metals						
05051	Anchor Bolts, Toggle Bolts, and Concrete Inserts	LS	1	$ 500.00	1.0	$500.00
05501	Miscellaneous Metal Fabrications (includes access platforms)	LS	1	$ 15,000.00	1.0	$15,000.00
	Division 5 Total					**$15,500.00**
Division 6 - Wood & Plastics						
Division 7 - Thermal & Moisture Protection						
07920	Caulking and Sealants	LS	1	$ 250.00	1.0	$250.00
	Division 7 Total					**$250.00**
Division 8 - Doors & Windows						
Division 9 - Finishes						
09611	Concrete Hardener	SF	3,600	$ 3.25	1.0	$11,700.00
09900	Painting	LS	1	$ 5,000.00	1.0	$5,000.00
	Division 9 Total					**$16,700.00**

Specification Section		Unit	Quantity	Cost	Installation Factor	Total
Division 10 - Specialties						
Division 11 - Equipment						
11197	Strong Base Anion Equipment (BWiX Treatment System)	LS	1	$ 683,824.00	1.0	$683,824.00
11530	Pumps, General	EA	2	$ 14,800.00	1.2	$35,520.00
	Division 11 Total					**$719,344.00**
Division 12 - Furnishings						
Division 13 - Special Construction						
13125	FRP Walk-In Enclosure (for MCC/electrical gear)	LS	1	$ 18,000.00	1.2	$21,600.00
	Division 13 Total					**$21,600.00**
Division 14 - Conveying Systems						
Division 15 - Mechanical						
15051	Buried Piping Installation					
	8" Influent Piping	FT	250	$ 21.00	1.4	$7,350.00
	Effluent Piping	FT	250	$ 21.00	1.4	$7,350.00
	Waste Piping	FT	100	$ 21.00	1.4	$2,940.00
15052	Exposed Piping Installation					
	Ion Exchange System Process Piping			In 11195		
15055	Pipe Hangers and Supports	LS	1	$ 5,000.00	1.0	$5,000.00
15061	Ductile Iron Pipe			In 15051		
15067	Thermoplastic Pipe			In 15051 & 15052		
15100	Valves, 4-Inch and Larger	LS	1	$ 10,000.00	1.2	$12,000.00
	Division 15 Total					**$34,640.00**
Division 16 - Electrical						
16050	General Provisions	LS	1	$ 150,000.00	1.0	$150,000.00
	Division 16 Total					**$150,000.00**
Division 17 - Instrumentation and Control						
17400	Instrumentation and Control	LS	1	$ 115,000.00	1.0	$120,000.00
	Division 17 Total					**$120,000.00**
				Division 1 - 17 Subtotal		**$1,276,224.00**

City of Glendale
Operation and Maintenance Costs - Strong Base Anion (SBA) - 500 gpm

Company:	Malcolm Pirnie, Inc.	**Date:**	29-Sep-06
Project:	Phase III Hexavalent Chromium Demonstration System	**Estimator:**	SMD
Submittal:	Conceptual Design Level Cost Estimate	**Checker:**	GB, MJM, NB
Work Task:		**Cost Index:**	ENR CCI = 8572.47 (Los Angeles, September 2006)

Item	Description	O&M Misc.	Energy	Chemicals	Waste Disposal	Total O&M	Notes
1.0	**Regenerable Strong Base Anion System**						
1.1	Salt Usage			$2,379		$2,379.03	Assumed 59-lbs per acre-foot of water at $100/ton
1.2	Pre-Filters filter replacement	$1,440				$1,440.00	Assumed monthly replacement of bag filters
1.3	Effluent Booster Pumps		$6,345			$6,344.53	Boosting IX effluent to aerator influent. 500-gpm @ 50' TDH
1.4	Miscellaneous Maintenance	$14,387				$14,386.88	Assumed 2% of equipment costs (Div 11)
	Subtotal - SBA System					**$24,550.44**	
2.0	**Waste Disposal**						
2.1	Backwash Waste Disposal (Non-Haz)				$696	$696.42	Based on backwash volume of 262,800 gallons per year disposed of at sanitary sewer.
2.2	Brine Waste Disposal (Haz)				$60,225	$60,225.00	Based on 1,110-gpd disposed off site to sewer leading to Hyperion (includes trucking costs) at $0.15 per gal.
2.3	Solids Waste Disposal				$2,817	$2,816.88	Based on 12 drums of solid waste per year at $445/ton.
	Subtotal - Waste Disposal					**$63,738.30**	
3.0	**Labor**						
3.1	Labor	$50,000				$50,000.00	Labor based on 0..5 FTE at $100,000/yr.
	Subtotal - Labor					**$50,000.00**	
TOTAL	**Subtotals**	$65,827	$6,345	$2,379	$63,738		
	Total					**$138,288.74**	

BasinWater Strong-Base Anion Exchange – 1,000 gpm:

City of Glendale
Capital Cost Estimate - Strong Base Anion (SBA) - 1000 gpm

Company: Malcolm Pirnie, Inc.
Project: Phase III Hexavalent Chromium Demonstration System
Submittal: Conceptual Design Level Cost Estimate
Work Task:

Date: 29-Sep-06
Estimator: SMD
Checker: GB, MJM, NB
Cost Index: ENR CCI = 8572.47 (Los Angeles, September 2006)

Division Summary		Total
Division 1 - General Conditions		$75,000.00
Division 2 - Site Construction		$81,690.00
Division 3 - Concrete		$66,500.00
Division 4 - Masonry		$0.00
Division 5 - Metals		$15,500.00
Division 6 - Wood & Plastics		$0.00
Division 7 - Thermal & Moisture Protection		$0.00
Division 8 - Doors & Windows		$0.00
Division 9 - Finishes		$16,700.00
Division 10 - Specialties		$0.00
Division 11 - Equipment		$848,702.00
Division 12 - Furnishings		$0.00
Division 13 - Special Construction		$21,600.00
Division 14 - Conveying Systems		$0.00
Division 15 - Mechanical		$50,000.00
Division 16 - Electrical		$210,000.00
Division 17 - Instrumentation and Control		$120,000.00
	Division 1 - 17 Subtotal	**$1,505,692.00**
	Insurance @ 2.5%	**$37,642.30**
	Bonds @ 2.0%	**$30,113.84**
	Overhead & Profit @ 10%	**$150,569.20**
	Engineering @ 10%	**$150,569.20**
	Total	**$1,874,586.54**
	Contingency @ 20.0%	**$374,917.31**
	CONCEPTUAL LEVEL PROBABLE CONSTRUCTION COST	**$2,249,503.85**
	(September 2006)	

City of Glendale
Capital Cost Estimate - Strong Base Anion (SBA) - 1000 gpm

Company: Malcolm Pirnie, Inc.
Project: Phase III Hexavalent Chromium Demonstration System
Submittal: Conceptual Design Level Cost Estimate
Work Task:

Date: 29-Sep-06
Estimator: SMD
Checker: GB, MJM, NB
Cost Index: ENR CCI = 8572.47 (Los Angeles, September 2006)

Specification Section		Unit	Quantity	Cost	Installation Factor	Total
Division 1 - General Conditions						
Div 1	General Conditions	LS	1	$ 75,000.00	1.0	$ 75,000.00
	Mobilization/Demobilization					
	Division 1 Total					**$75,000.00**
Division 2 - Site Construction						
02220	Site Preparation	LS	1	$ 5,000.00	1.0	$5,000.00
02230	Clearing					
02315	Excavation and Backfill					
	Excavation and grading of of demonstration facility limits	LS	1	$ 50,000.00	1.0	$50,000.00
	Excess Material to be removed	CY	1,570	$ 8.00	1.0	$12,560.00
	Backfill	CY	785	$ 18.00	1.0	$14,130.00
	Division 2 Total					**$81,690.00**
Division 3 - Concrete						
03100	Concrete Formwork			In 03300		
03300	Concrete					
	Equipment Slab	CY	133	$ 500.00	1.0	$66,500.00
	Division 3 Total					**$66,500.00**
Division 4 - Masonry						
Division 5 - Metals						
05051	Anchor Bolts, Toggle Bolts, and Concrete Inserts	LS	1	$ 500.00	1.0	$500.00
05501	Miscellaneous Metal Fabrications (includes access platforms)	LS	1	$ 15,000.00	1.0	$15,000.00
	Division 5 Total					**$15,500.00**
Division 6 - Wood & Plastics						
Division 7 - Thermal & Moisture Protection						
Division 8 - Doors & Windows						
Division 9 - Finishes						
09611	Concrete Hardener	SF	3,600	$ 3.25	1.0	$11,700.00
09900	Painting	LS	1	$ 5,000.00	1.0	$5,000.00
	Division 9 Total					**$16,700.00**

Specification Section		Unit	Quantity	Cost	Installation Factor	Total
Division 10 - Specialties						
Division 11 - Equipment						
11197	Strong Base Anion Equipment (BWiX Treatment System)	LS	1	$ 804,782.00	1.0	$804,782.00
11530	Pumps, General	EA	2	$ 18,300.00	1.2	$43,920.00
	Division 11 Total					**$848,702.00**
Division 12 - Furnishings						
Division 13 - Special Construction						
13125	FRP Walk-In Enclosure (for MCC/electrical gear)	LS	1	$ 18,000.00	1.2	$21,600.00
	Division 13 Total					**$21,600.00**
Division 14 - Conveying Systems						
Division 15 - Mechanical						
15051	Buried Piping Installation					
	8" Influent Piping	FT	250	$ 25.00	1.4	$8,750.00
	Effluent Piping	FT	250	$ 25.00	1.4	$8,750.00
	Brine Waste Piping	FT	100	$ 25.00	1.4	$3,500.00
15052	Exposed Piping Installation					
	Ion Exchange System Process Piping			In 11195		
15055	Pipe Hangers and Supports	LS	1	$ 5,000.00	1.0	$5,000.00
15061	Ductile Iron Pipe			In 15051		
15067	Thermoplastic Pipe			In 15051 & 15052		
15100	Valves, 4-Inch and Larger	LS	1	$ 20,000.00	1.2	$24,000.00
	Division 15 Total					**$50,000.00**
Division 16 - Electrical						
16050	General Provisions	LS	1	$ 210,000.00	1.0	$210,000.00
	Division 16 Total					**$210,000.00**
Division 17 - Instrumentation and Control						
17400	Instrumentation and Control	LS	1	$ 165,000.00	1.0	$120,000.00
	Division 17 Total					**$120,000.00**
					Division 1 - 17 Subtotal	**$1,505,692.00**

City of Glendale
Operation and Maintenance Costs - Strong Base Anion (SBA) - 1000 gpm

Company:	Malcolm Pirnie, Inc.	**Date:**	29-Sep-06
Project:	Phase III Hexavalent Chromium Demonstration System	**Estimator:**	SMD
Submittal:	Conceptual Design Level Cost Estimate	**Checker:**	GB, MJM, NB
Work Task:		**Cost Index:**	ENR CCI = 8572.47 (Los Angeles, September 2006)

Item	Description	O&M Misc.	Energy	Chemicals	Waste Disposal	Total O&M	Notes
1.0	**Regenerable Strong Base Anion System**						
1.1	Salt Usage			$4,758		$4,758.06	Assumed 59-lbs per acre-foot of water at $100/ton
1.2	Pre-Filters filter replacement	$1,440				$1,440.00	Assumed monthly replacement of bag filters
1.3	Effluent Booster Pumps		$6,345			$6,344.53	Boosting IX effluent to aerator influent. 1,000-gpm @ 50' TDH
1.4	Miscellaneous Maintenance	$16,974				$16,974.04	Assumed 2% of equipment costs (Div 11)
	Subtotal - SBA System					**$29,516.63**	
2.0	**Waste Disposal**						
2.1	Backwash Waste Disposal (Non-Haz)				$1,393	$1,392.84	Based on backwash volume of 525,600 gallons per year disposed of at sanitary sewer.
2.2	Brine Waste Disposal (Haz)				$120,450	$120,450.00	Based on 2,200-gpd disposed off site (includes trucking costs) at $0.15 per gal.
2.3	Solids Waste Disposal				$5,634	$5,633.75	Based on 24 drums of solid waste per year at $445/ton.
	Subtotal - Waste Disposal					**$1,392.84**	
3.0	**Labor**						
3.1	Labor	$50,000				$50,000.00	Labor based on 0.5 FTE at $100,000/yr.
	Subtotal - Labor					**$50,000.00**	
TOTAL	**Subtotals**	$68,414	$6,345	$4,758	$127,477		
	Total					**$206,993.23**	

Reduction/Coagulation/Filtration – 500 gpm:

City of Glendale
Capital Cost Estimate - Reduction, Coagulation, Filtration (RCF) - 500 gpm

Company: Malcolm Pirnie, Inc.
Project: Phase III Hexavalent Chromium Demonstration System
Submittal: Conceptual Design Level Cost Estimate
Work Task:

Date: 29-Sep-06
Estimator: SMD
Checker: GB, MJM, NB
Cost Index: ENR CCI = 8572.47 (Los Angeles, September 2006)

Division Summary		Total
Division 1 - General Conditions		$50,000.00
Division 2 - Site Construction		$81,690.00
Division 3 - Concrete		$92,500.00
Division 4 - Masonry		$0.00
Division 5 - Metals		$500.00
Division 6 - Wood & Plastics		$0.00
Division 7 - Thermal & Moisture Protection		$0.00
Division 8 - Doors & Windows		$0.00
Division 9 - Finishes		$21,250.00
Division 10 - Specialties		$500.00
Division 11 - Equipment		$1,337,400.00
Division 12 - Furnishings		$0.00
Division 13 - Special Construction		$21,600.00
Division 14 - Conveying Systems		$0.00
Division 15 - Mechanical		$39,640.00
Division 16 - Electrical		$150,000.00
Division 17 - Instrumentation and Control		$115,000.00
	Division 1 - 17 Subtotal	**$1,910,080.00**
	Insurance @ 2.5%	**$47,752.00**
	Bonds @ 2.0%	**$38,201.60**
	Overhead & Profit @ 10%	**$191,008.00**
	Engineering @ 10%	**$191,008.00**
	Total	**$2,378,049.60**
	Contingency @ 20.0%	**$475,609.92**
	CONCEPTUAL LEVEL PROBABLE CONSTRUCTION COST	**$2,853,659.52**
	(September 2006)	

City of Glendale
Capital Cost Estimate - Reduction, Coagulation, Filtration (RCF) - 500 gpm

Company: Malcolm Pirnie, Inc.
Project: Phase III Hexavalent Chromium Demonstration System
Submittal: Conceptual Design Level Cost Estimate
Work Task:

Date: 29-Sep-06
Estimator: SMD
Checker: GB, MJM, NB
Cost Index: ENR CCI = 8572.47 (Los Angeles, September 2006)

	Specification Section	Unit	Quantity	Cost	Installation Factor	Total
Division 1 - General Conditions						
Div 1	General Conditions	LS	1	$ 50,000.00	1.0	$50,000.00
	Mobilization/Demobilization					
	Division 1 Total					**$50,000.00**
Division 2 - Site Construction						
02220	Site Preparation	LS	1	$ 5,000.00	1.0	$5,000.00
02230	Clearing					
02315	Excavation and Backfill					
	Excavation and grading of of demonstration facility limits	LS	1	$ 50,000.00	1.0	$50,000.00
	Excess Material to be removed	CY	1,570	$ 8.00	1.0	$12,560.00
	Backfill	CY	785	$ 18.00	1.0	$14,130.00
	Division 2 Total					**$81,690.00**
Division 3 - Concrete						
03100	Concrete Formwork			In 03300		
03300	Concrete					
	Equipment Slab (5,000-sf)	CY	185	$ 500.00	1.0	$92,500.00
	Division 3 Total					**$92,500.00**
Division 4 - Masonry						
Division 5 - Metals						
05051	Anchor Bolts, Toggle Bolts, and Concrete Inserts	LS	1	$ 500.00	1.0	$500.00
	Division 5 Total					**$500.00**
Division 6 - Wood & Plastics						
Division 7 - Thermal & Moisture Protection						
Division 8 - Doors & Windows						
Division 9 - Finishes						
09611	Concrete Hardener	SF	5,000	$ 3.25	1.0	$16,250.00
09900	Painting	LS	1	$ 5,000.00	1.0	$5,000.00
	Division 9 Total					**$21,250.00**
Division 10 - Specialties						
10400	Identification Devices	LS	1	$ 500.00	1.0	$500.00
	Division 10 Total					**$500.00**

	Specification Section	Unit	Quantity	Cost	Installation Factor	Total
Division 11 - Equipment						
11179	Fiberglass Reinforced Plastic Tanks (Ferrous Sulfate Bulk Storage)	LS	1	$ 5,000.00	1.2	$6,000.00
11180	Backwash Water Holding Tank with mixer (20,000-gal.)	LS	1	$ 36,600.00	1.2	$43,920.00
11218	Ferrous Sulfate Feed System	LS	1	$ 30,000.00	1.2	$36,000.00
11219	*Polymer Addition System*	*LS*	*1*	*$ 20,000.00*	1.2	$24,000.00
11216	Gravity Settler	EA	1	$ 82,900.00	1.2	$99,480.00
11217	Dual Media Filter (5-cell)	EA	1	$ 450,000.00	1.2	$540,000.00
11311	Reduction Tank (30,000-gal. Steel Tank) w/ mixers	LS	1	$ 70,000.00	1.2	$84,000.00
11312	Aeration Chamber	LS	1	$ 65,000.00	1.2	$78,000.00
	Steel Tank (10,000-gallons)					
	Blowers					
	Coarse Air Diffusers					
	Air Compressor					
11315	Belt Filter Press (1-meter)	EA	1	$ 220,000.00	1.2	$264,000.00
	Backwash Water Storage Tank (30,000-gal.)	LS	1	$ 45,000.00	1.2	$54,000.00
15130	Pumps, Booster	EA	6	$ 15,000.00	1.2	$108,000.00
	Division 11 Total					**$1,337,400.00**
Division 12 - Furnishings						
Division 13 - Special Construction						
13125	FRP Walk-In Enclosure	LS	1	$ 18,000.00	1.2	$21,600.00
	Division 13 Total					**$21,600.00**
Division 14 - Conveying Systems						
Division 15 - Mechanical						
15051	Buried Piping Installation					
	8" Influent Piping	FT	250	$ 21.00	1.4	$7,350.00
	Effluent Piping	FT	250	$ 21.00	1.4	$7,350.00
	Backwash Waste Piping	FT	100	$ 21.00	1.4	$2,940.00
15052	Exposed Piping Installation					
	12" Influent Piping	FT		$ 20.00	1.4	$0.00
15055	Pipe Hangers and Supports	LS	1	$ 10,000.00	1.0	$10,000.00
15061	Ductile Iron Pipe			In 15051		
15067	Thermoplastic Pipe			In 15051 & 15052		
15100	Valves & Fittings (Misc.), 4-Inch and Larger	LS	0	$ 10,000.00	1.2	$12,000.00
	Division 15 Total					**$39,640.00**
Division 16 - Electrical						
16050	General Provisions	LS	1	$ 150,000.00	1.0	$150,000.00
	Division 16 Total					**$150,000.00**
Division 17 - Instrumentation and Control						
17400	Instrumentation and Control	LS	1	$ 115,000.00	1.0	$115,000.00
	Division 17 Total					**$115,000.00**
					Division 1 - 17 Subtotal	**$1,910,080.00**

City of Glendale

Operation and Maintenance Costs - Reduction, Coagulation, Filtration (RCF) - 500 gpm

Company:	Malcolm Pirnie, Inc.	**Date:**	29-Sep-06
Project:	Phase III Hexavalent Chromium Demonstration System	**Estimator:**	SMD
Submittal:	Conceptual Design Level Cost Estimate	**Checker:**	GB, MJM, NB
Work Task:		**Cost Index:**	ENR CCI = 8572.47 (Los Angeles, September 2006)

Item	Description	O&M Misc	Energy	Chemicals	Waste Disposal	Total O&M	Notes
1.0	**Reduction Coagulation Filtration**						
1.1	Media Replacement	$10,000				$10,000.00	Assumed 10% media loss per year.
1.2	Booster/Backwash Pumps/Blowers		$35,928			$35,927.83	Booster pumps, backwash pumps and aeration blowers.
1.3	Chemicals			$7,884		$7,884.00	Assuming a 25:1 Fe:Cr mass ratio, Fe dose of 2.5 mg/L as Fe, and cost of 5% (as Fe) ferrous sulfate at $0.60/gallon
1.4	Miscellaneous Maintenance	$26,748				$26,748.00	Assumed 2% of equipment costs (Div 11)
	Subtotal - RCF System					**$80,559.83**	
2.0	**Waste Disposal**						
2.1	Liquid Waste Disposal (Non-Hazardous)				$781	$781.00	Backwash water volume of 4% of water treated; 0.58% of backwash water as settled sludge, 99.42% recycled; 3% solids in settled sludge; 80% filter press dewatering efficiency; total backwash water to sewer (in Glendale) of 28,000 gpd waste (for 500-gpm system). If no backwash water could be recycled, this annual sewer disposal cost would be $19,534 in Glendale and $33,895 in LA
2.2	Residuals Disposal (Hazardous)				$22,695	$22,695.00	Residuals disposal based on 20 tons/year at $445/ton
	Subtotal - Waste Disposal					**$781.00**	
3.0	**Labor**						
3.1	Labor	$50,000				$50,000.00	Labor based on 0.5 FTE at $100,000/yr.
	Subtotal - Labor					**$50,000.00**	
TOTAL	**Subtotals**	$86,748	$35,928	$7,884	$23,476		
	Total					**$154,035.83**	

Reduction/Coagulation/Filtration – 1,000 gpm:

City of Glendale
Capital Cost Estimate - Reduction, Coagulation, Filtration (RCF) - 1000 gpm

Company: Malcolm Pirnie, Inc.
Project: Phase III Hexavalent Chromium Demonstration System
Submittal: Conceptual Design Level Cost Estimate
Work Task:

Date: 29-Sep-06
Estimator: SMD
Checker: GB, MJM, NB
Cost Index: ENR CCI = 8572.47 (Los Angeles, September 2006)

Division Summary		Total
Division 1 - General Conditions		$75,000.00
Division 2 - Site Construction		$81,690.00
Division 3 - Concrete		$111,000.00
Division 4 - Masonry		$0.00
Division 5 - Metals		$500.00
Division 6 - Wood & Plastics		$0.00
Division 7 - Thermal & Moisture Protection		$0.00
Division 8 - Doors & Windows		$0.00
Division 9 - Finishes		$24,500.00
Division 10 - Specialties		$500.00
Division 11 - Equipment		$1,654,560.00
Division 12 - Furnishings		$0.00
Division 13 - Special Construction		$21,600.00
Division 14 - Conveying Systems		$0.00
Division 15 - Mechanical		$55,000.00
Division 16 - Electrical		$210,000.00
Division 17 - Instrumentation and Control		$130,000.00
	Division 1 - 17 Subtotal	**$2,364,350.00**
	Insurance @ 2.5%	**$59,108.75**
	Bonds @ 2.0%	**$47,287.00**
	Overhead & Profit @ 10%	**$236,435.00**
	Engineering @ 10%	**$236,435.00**
	Total	**$2,943,615.75**
	Contingency @ 20.0%	**$588,723.15**
	CONCEPTUAL LEVEL PROBABLE CONSTRUCTION COST	**$3,532,338.90**
	(September 2006)	

City of Glendale
Capital Cost Estimate - Reduction, Coagulation, Filtration (RCF) - 1000 gpm

Company: Malcolm Pirnie, Inc.
Project: Phase III Hexavalent Chromium Demonstration System
Submittal: Conceptual Design Level Cost Estimate
Work Task:

Date: 29-Sep-06
Estimator: SMD
Checker: GB, MJM, NB
Cost Index: ENR CCI = 8572.47 (Los Angeles, September 2006)

	Specification Section	Unit	Quantity	Cost	Installation Factor	Total
Division 1 - General Conditions						
Div 1	General Conditions	LS	1	$ 75,000.00	1.0	$75,000.00
	Mobilization/Demobilization					
	Division 1 Total					**$75,000.00**
Division 2 - Site Construction						
02220	Site Preparation	LS	1	$ 5,000.00	1.0	$5,000.00
02230	Clearing					
02315	Excavation and Backfill					
	Excavation and grading of of demonstration facility limits	LS	1	$ 50,000.00	1.0	$50,000.00
	Excess Material to be removed	CY	1,570	$ 8.00	1.0	$12,560.00
	Backfill	CY	785	$ 18.00	1.0	$14,130.00
	Division 2 Total					**$81,690.00**
Division 3 - Concrete						
03100	Concrete Formwork			In 03300		
03300	Concrete					
	Equipment Slab (3,500-sf)	CY	222	$ 500.00	1.0	$111,000.00
	Division 3 Total					**$111,000.00**
Division 4 - Masonry						
Division 5 - Metals						
05051	Anchor Bolts, Toggle Bolts, and Concrete Inserts	LS	1	$ 500.00	1.0	$500.00
	Division 5 Total					**$500.00**
Division 6 - Wood & Plastics						
Division 7 - Thermal & Moisture Protection						
Division 8 - Doors & Windows						
Division 9 - Finishes						
09611	Concrete Hardener	SF	6,000	$ 3.25	1.0	$19,500.00
09900	Painting	LS	1	$ 5,000.00	1.0	$5,000.00
	Division 9 Total					**$24,500.00**
Division 10 - Specialties						
10400	Identification Devices	LS	1	$ 500.00	1.0	$500.00
	Division 10 Total					**$500.00**

Specification Section		Unit	Quantity	Cost	Installation Factor	Total
Division 11 - Equipment						
11179	Fiberglass Reinforced Plastic Tanks (Ferrous Sulfate Bulk Storage)	LS	1	$ 5,000.00	1.2	$6,000.00
11180	Backwash Water Holding Tank with mixer (20,000-gal.)	LS	1	$ 36,600.00	1.2	$43,920.00
11218	Ferrous Sulfate Feed System	LS	1	$ 34,500.00	1.2	$41,400.00
11216	Gravity Settler	EA	1	$ 82,900.00	1.2	$99,480.00
11217	Dual Media Filter (5-cell)	EA	1	$ 625,000.00	1.2	$750,000.00
11219	Polymer Addition System	LS	1	$ 20,000.00	1.2	$24,000.00
11311	Reduction Tank (60,000-gal. Steel Tank) w/ mixers	LS	1	$ 110,000.00	1.2	$132,000.00
11312	Aeration Chamber	LS	1	$ 90,000.00	1.2	$108,000.00
	Steel Tank (10,000-gallons)					
	Blowers					
	Coarse Air Diffusers					
	Air Compressor					
11315	Belt Filter Press (1-meter)	EA	1	$ 220,000.00	1.2	$264,000.00
	Backwash Water Storage Tank (30,000-gal.)	LS	1	$ 45,000.00	1.2	$54,000.00
15130	Pumps, Booster	EA	6	$ 18,300.00	1.2	$131,760.00
	Division 11 Total					**$1,654,560.00**
Division 12 - Furnishings						
Division 13 - Special Construction						
13125	FRP Walk-In Enclosure (MCC)	LS	1	$ 18,000.00	1.2	$21,600.00
	Division 13 Total					**$21,600.00**
Division 14 - Conveying Systems						
Division 15 - Mechanical						
15051	Buried Piping Installation					
	12" Influent Piping	FT	250	$ 25.00	1.4	$8,750.00
	Effluent Piping	FT	250	$ 25.00	1.4	$8,750.00
	Backwash Waste Piping	FT	100	$ 25.00	1.4	$3,500.00
15052	Exposed Piping Installation					
	12" Influent Piping	FT		$ 21.00	1.4	$0.00
15055	Pipe Hangers and Supports	LS	1	$ 10,000.00	1.0	$10,000.00
15061	Ductile Iron Pipe			In 15051		
15067	Thermoplastic Pipe			In 15051 & 15052		
15100	Valves & Fittings (Misc.), 4-Inch and Larger	LS	0	$ 20,000.00	1.2	$24,000.00
	Division 15 Total					**$55,000.00**
Division 16 - Electrical						
16050	General Provisions	LS	1	$ 210,000.00	1.0	$210,000.00
	Division 16 Total					**$210,000.00**
Division 17 - Instrumentation and Control						
17400	Instrumentation and Control	LS	1	$ 130,000.00	1.0	$130,000.00
	Division 17 Total					**$130,000.00**
				Division 1 - 17 Subtotal		**$2,364,350.00**

City of Glendale

Operation and Maintenance Costs - Reduction, Coagulation, Filtration (RCF) - 1000 gpm

Company:	Malcolm Pirnie, Inc.	**Date:**	29-Sep-06
Project:	Phase III Hexavalent Chromium Demonstration System	**Estimator:**	SMD
Submittal:	Conceptual Design Level Cost Estimate	**Checker:**	GB, MJM, NB
Work Task:		**Cost Index:**	ENR CCI = 8572.47 (Los Angeles, September 2006)

Item	Description	O&M Misc.	Energy	Chemicals	Waste Disposal	Total O&M	Notes
1.0	**Reduction Coagulation Filtration**						
1.1	Media Replacement	$10,000				$10,000.00	Assumed 10% media loss per year.
1.2	Booster/Backwash Pumps/Blowers		$35,928			$35,927.83	Booster pumps, backwash pumps and aeration blowers.
1.3	Chemicals			$15,768		$15,768.00	Assuming a 25:1 Fe:Cr mass ratio, Fe dose of 2.5 mg/L as Fe, and cost of 5% (as Fe) ferrous sulfate at $0.60/gallon ; Polymer addition at 1 mg/L at $
1.4	Miscellaneous Maintenance	$33,091				$33,091.20	Assumed 2% of equipment costs (Div 11)
	Subtotal - RCF System					**$94,787.03**	
2.0	**Waste Disposal**						
2.1	Liquid Waste Disposal (Non-Hazardous)				$2,712	$2,712.00	Backwash water volume of 4% of water treated; 0.58% of backwash water as settled sludge, 99.42% recycled; 3% solids in settled sludge; 80% filter press dewatering efficiency; total backwash water to sewer in Glendale of 57,600 gpd waste (for 1,000-gpm system). If no backwash water could be recycled, this annual sewer disposal cost would be $39,069 in Glendale and $67,789 in LA
2.2	Residuals Disposal (Hazardous)				$45,390	$45,390.00	Residuals disposal based on 40 tons/year at $445/ton
	Subtotal - Waste Disposal					**$2,712.00**	
3.0	**Labor**						
3.1	Labor	$50,000				$50,000.00	Labor based on 0.5 FTE at $100,000/yr.
	Subtotal - Labor					**$50,000.00**	
TOTAL	**Subtotals**	$93,091	$35,928	$15,768	$48,102		
	Total					**$192,889.03**	

REFERENCES

AACE (American Association of Cost Engineers). 2000. AACE International Recommended Practice No. 18R-97, Cost Estimate Classification System. Morgantown, WV: AACE.

Bahowick, S., D. Dobie, and G. Kumamoto. 1996. Ion-Exchange Resin for Removing Hexavalent Chromium from Groundwater at Treatment Facilities. UCRL-JC-122285 Abs. Lawrence, Calif.: Lawrence Livermore National Laboratory.

BasinWater. 2006. Proposed Cost Information for Cr(VI) Treatment. Submitted to McGuire Malcolm Pirnie. September 2006.

Berry, A.J., and H.S.C. O'Neill. 2004. A XANES Determination of the Oxidation State of Chromium in Silicate Glasses. *Am. Mineralogist*, 89(5–6):790–798.

Blute, N.K., M.J. McGuire, C. Seidel, and J. Byerrum. No Free Lunch: Perchlorate Removal and Resin Byproducts. Presented at the AWWA Annual Conference and Exhibition, San Antonio, Texas, June 12–15.

Bond, D.L., and S. Fendorf. 2003. Kinetics and Structural Constraints of Chromate Reduction by Green Rusts. *Environ. Sci. Technol.*, 37(12):2750–2757.

Brandhuber, P., M. Frey, M.J. McGuire, P. Chao, C. Seidel, G. Amy, J. Yoon, L. McNeill, and K. Banerjee. 2004. *Treatment Options for Low-Level Hexavalent Chromium Removal Tested at Bench Scale*. Denver, Colo.: AwwaRF.

Clifford, D.A. 1999. Ion Exchange and Inorganic Adsorption. In *Water Quality and Treatment: A Handbook of Community Water Supplies*. Edited by R.D. Letterman. New York: McGraw Hill.

Cornwell, D.A. 1999. Water Treatment Plant Residuals Management. In *Water Quality and Treatment: A Handbook of Community Water Supplies*. Edited by R.D. Letterman. New York: McGraw Hill.

Höll, W.H., C. Bartosch, X. Zhao, and S. He. 2002. Elimination of Trace Heavy Metals from Drinking Water Using Weakly Basic Anion Exchangers. *Geo- und Wassertechnologie*, 1(3):5–20.

Horng, L.L., and D.A. Clifford. 1997. The Behavior of Polyprotic Anions in Ion Exchange Resins. *Reactive and Functional Polymers*, 35(1/2):41–54.

Jacobs, J.A., and S.M. Testa. 2005. Overview of Chromium(VI) in the Environment. In *Chromium(VI) Handbook*, Edited by J. Guertin, J.A. Jacobs, and C.P. Avakian. New York: CRC Press.

Jakobsen, K., and R. Laska. 1977. Advanced Treatment Methods for Electroplating Wasters. *Pollution Eng.*, 8:42–46.

McGuire, M.J., N.K. Blute, C. Seidel, G. Qin, and L. Fong. 2006. Pilot-Scale Studies of Hexavalent Chromium Removal From Drinking Water. *Jour. AWWA*, 98(2):134–143.

MEC (McGuire Environmental Consultants, Inc.). 2005. The Treatment of Hexavalent Chromium (Cr(VI)) in the City of Glendale Groundwater Supply: Phase II Demonstration of Pilot-Scale Treatment Technologies. Final report submitted to the City of Glendale, Calif.

Najm, I., and R.R. Trussell. 2001. NDMA Formation in Water and Wastewater. *Jour. AWWA*, 93(2):92–99.

O'Day, P.A., S.A. Carroll, S. Randall, R.E. Martinelli, S.L. Anderson, J. Jelinski, and J.P. Knezovich. 2000. Metal Speciation and Bioavailability in Contaminated Estuary Sediments, Alameda Naval Air Station, California. *Environ. Sci. Technol.*, 34(17):3665–3673.

Office of Administrative Law. 1991. Title 22, California Code of Regulations, Section 66261.126, Appendix II [Online]. Available: <http://oal.ca.gov/linkedslice/default.asp?SP=CCR-1000&Action=welcome>. [cited August 30, 2007]

Patterson, J. 1985. *Industrial Wastewater Treatment Technology.* 2nd ed. Boston: Butterworth.

Peschman, T. 2006. Personal communication.

Qin, G., M.J. McGuire, N.K. Blute, C.J. Seidel, and L. Fong. 2005. Hexavalent Chromium Removal by Reduction With Ferrous Sulfate, Coagulation, and Filtration: A Pilot-Scale Study. *Environ. Sci. Technol.,* 39(16):6321–6327.

Ravel, B.; Newville, M. 2005. ATHENA, ARTEMIS, HEPHAESTUS: Data Analysis for X-Ray Absorption Spectroscopy using IFEFFIT. *J. Synchrotron Rad.* 12:537–541.

Richardson, E., E. Stobbe, and S. Bernstein. 1968. Ion Exchange Traps Chromates for Reuse. *Environ. Sci. Technol.,* 2:1006–1016.

Sengupta, A.K., and D. Clifford. 1986. Important Process Variables in Chromate Ion Exchange. *Environ. Sci. Technol.,* 20(2):149–155.

Sengupta, A.K., D. Clifford, and S. Subramonian. 1986. Chromate Ion-Exchange Process at Alkaline pH. *Wat. Res.,* 20(9):1177–1184.

USEPA (U.S. Environmental Protection Agency). 1994. Method 200.8: Determination of Trace Elements in Waters and Wastes by Inductively Coupled Plasma – Mass Spectrometry. EPA-600/R-94/111.

USEPA (U.S. Environmental Protection Agency). 1996. Method 1636: Determination of Hexavalent Chromium by Ion Chromatography. EPA-821/R-96/003.

USEPA (U.S. Environmental Protection Agency). 1998. Test Methods for Evaluating Solid Waste, Physical/Chemical Methods. Publication SW-846.

Siemens. 2006a. Proposed Cost Information for Hexavalent Chromium Treatment. Submitted to McGuire Malcolm Pirnie, September 2006.

Siemens. 2006b. Proposed Retrofit Cost Information for Hexavalent Chromium Treatment at GS-3. Submitted to McGuire Malcolm Pirnie, March 2006.

Wilkin, R.T., C. Su, R.G. Ford, and C.J. Paul. 2005. Chromium-Removal Processes During Groundwater Remediation by a Zerovalent Iron Permeable Reactive Barrier. *Environ. Sci. Technol.,* 39(12):4599–4605.

ABBREVIATIONS

AACE	American Association of Cost Engineers
AF	acre foot
ASTM	American Society of Testing and Materials
AwwaRF	Awwa Research Foundation
BSE	backscatter electron
BV	bed volume
cf	cubic foot
$CaCO_3$	calcium carbonate
Cr(III)	trivalent chromium
Cr(VI)	hexavalent chromium
CrO_4^{2-}	chromate
Cr_2O_3	chromium oxide
$Cr_2O_7^{2-}$	dichromate
$Cr(OH)_3$	chromium (III) hydroxide
CU	University of Colorado at Boulder
DHS	Department of Health Services
EBCT	empty bed contact time
eV	electron volt
Fe^{2+}	ferrous iron, iron (II)
Fe^{3+}	ferric iron, iron (III)
FRP	fiberglass reinforced plastic
FTE	full-time equivalent
g	gram
GAC	granular activated carbon
GN	Glendale North
gpm	gallon per minute
gpm/ft^2	gallon per minute per square foot
GS	Glendale South
GWTP	Glendale Water Treatment Plant
HCl	hydrochloric acid
$HCrO_4^-$	bichromate
HDPE	high density polyethylene
IC	ion chromatograph
ICP-MS	inductively coupled plasma–mass spectrometer

KPA	kinetic phosphorescence analysis
L	liter
LADWP	Los Angeles Department of Water and Power
lb	pounds
MCL	maximum contaminant level
MEC	McGuire Environmental Consultants, Inc.
mg/L	milligram per liter
min	minute
mmoles/g	millimoles per gram
MWDSC	Metropolitan Water District of Southern California
µg/g	microgram per gram
µg/L	microgram per liter
µm	micrometer
µS/cm	microsiemens per centimeter
NDBA	N-Nitrosodi-n-butylamine
NDEA	N-Nitrosodiethylamine
NDMA	N-Nitrosodimethylamine
NDPA	N-Nitrosodi-n-propylamine
NDPhA	N-Nitrosodiphenylamine
ng/L	nanogram per liter
NMEA	N-Nitrosomethylethylamine
NPIP	N-Nitrosopiperidine
NTU	Nephelometric Turbidity Unit
NWRI	National Water Research Institute
NYPR	N-Nitrosopyrrolidine
OEHHA	Office of Environmental Health Hazard Assessment
O&M	operations and maintenance
PAC	Project Advisory Committee
pCi/L	pico-curie per liter
PHG	public health goal
PO_4	phosphate ion
ppb	parts per billion
PVC	polyvinyl chloride
RCF	reduction/coagulation/filtration
redox	oxidation-reduction
R&H	Rohm & Haas
SBA	strong-base anion exchange
SEM	scanning electron microscopy
SiO_2	silica

SO_4	sulfate ion
STLC	soluble threshold limit concentration
TC	tailored collaboration
TCLP	Toxicity Characteristic Leaching Procedure
UCLA	University of California Los Angeles
UCMR2	Unregulated Contaminant Monitoring Regulation
UNESCO	United Nations Educational, Scientific and Cultural Organization
USEPA	U.S. Environmental Protection Agency
USU	Utah State University
VOC	volatile organic compound
WBA	weak-base anion exchange
WET	Waste Extraction Test
XANES	x-ray absorption near-edge structure
XRD	x-ray diffraction
XRF	x-ray fluorescence spectrometry

IWA Publishing's authorised EU representative for General Product Safety Regulations is Diane D'Arras, 15 rue Duret, 75116 Paris, France, e-mail: safety@iwap.co.uk.

Printed and bound by CPI Group (UK) Ltd, Croydon, CR0 4YY
17/09/2026
02218811-0013